Student's Guide to Analog Systems.
State Equations, Transforms, Convolution,
Controllability and Observability.

Dwight F. Mix

Professor Emeritus

University of Arkansas

Technical LAP series: Volume 6

Copyright 2015

MixPress.net

Contents

Objectives for Student's Guide to Analog Systems

What you should be able to do after completing each chapter.

Chap. 1. Linear Time Invariant (LTI) Systems

1. Describe (define) a linear system and apply this definition to test a given system for linearity.
2. Describe (define) a time-invariant system and apply this definition to test a given system for time invariance.
3. Given one input-output pair, use the LTI properties to find the response to a second input, where the second input is simply related to the first input.

Chap. 2. Formulation Procedures – How to derive the state model for a circuit

1. Write the state equations for a given electric circuit.

Chap. 3. The Solution of State Equations – Type A Networks

1. Find a closed form solution of the state equations for a type A network.

Chap. 4. The Solution of State Equations – Type B Networks

1. Find a closed form analytical solution for type B networks by impulse matching.
2. Find a closed form analytical solution for type B networks by reducing the form to type A equations.

Chap. 5. Approximate Digital Solution

1. Write a computer program to solve approximately the state equations for both type A and type B networks.

Chap. 6. Transfer Function

1. Describe (define) transfer function.
2. Find the transfer function $H(s)$ for a given LTI circuit.
3. Find the steady state response of an LTI circuit to an eternal exponential signal.

Chap. 7. Fourier Series

1. Describe (define) vector.
2. Calculate the inner product of two given waveforms.
3. Determine (select) which functions have a Fourier series.
4. Calculate and plot the Fourier series for a given waveform.
5. Calculate and plot the waveform for a given Fourier series.

Chap. 8. Response of LTI systems by Fourier series

1. Calculate the response of an LTI system to a signal expressed by its Fourier series.
2. Find and plot the power spectral density function for a periodic waveform.
3. Calculate the average power on a one ohm basis in the output signal of an LTI system, where the input signal is periodic.

Chap. 9. Fourier Transform

1. Determine (select) which functions have a Fourier transform.
2. Find and plot the Fourier transform for a given aperiodic time function.
3. Find and plot the time function corresponding to a given Fourier transform.

Chap. 10. Response of LTI Systems by Fourier Transform

1. Calculate the response of an LTI system to a signal expressed by its Fourier transform.
2. Find and plot the energy spectral density function for an energy signal.
3. Calculate the total energy on a one ohm basis in the output signal of an LTI system, where the input signal is an energy signal

Chap. 11. Convolution

1. Find the impulse response of given LTI circuits.
2. Convolve two given functions.
3. Find the response of LTI circuits to given input signals by convolution.

Chap. 12. Properties of Fourier Transform

1. Use the differentiation, delay, modulation, convolution, and multiplication properties of Fourier transforms to evaluate transforms of given functions.
2. Find bounds on the spectrum of a given time function by using the concepts of content, variation, and wiggliness.
3. Use the Paley-Wiener theorem to test the magnitude of transfer functions for physical realizability,

Chap. 13. Laplace Transform

1. Determine (select) which functions have a Laplace transform.
2. Find the Laplace transform for a given time function.
3. Find the time function corresponding to a given Laplace transform.

Chap. 14. Response of LTI Systems by Laplace Transform

1. Use partial fraction expansion to express the ratio of polynomials as the sum of partial fractions.
2. Find the response of an LTI system to an input signal that is expressed by its Laplace transform.

Chap. 15. Equivalence of System Models

1. Define controllability and observability,
2. Determine (select) which systems are controllable and which are observable.
3. Derive the equivalent system models (state model, transfer function, and impulse response) for given controllable and observable systems.

What is a LAP?

In 1972 I taught a Junior-level system theory course using an individualized approach. There were no lectures. Instead a special room was open every afternoon for the students to use. An instructor was present to help the students during the time when the room was open.

The students were required to complete one Learning Activity Package (LAP) per week. There were checkpoints throughout the semester. The student had to complete a certain portion of the course at each checkpoint by passing exams with no error over each LAP. If they failed to meet this requirement they were dropped from the course.

All but a few completed the course and received a grade of A. The course covered less material than normal, but the most common comment by the students on a questionnaire was that too much material was in the course. They had to master the material, so they earned their grade.

Most students learned more than normal, but the course was never taught again. Why? Because it took too many resources. A laboratory room was devoted to the course, and it took a professor and two graduate students to man the room each afternoon. I guess this proves the old adage, "There ain't no such thing as a free meal."

This text is the one used in that course. Each LAP has clearly defined objectives, and a self-test completes each objective.

Chapter 1

Linear Time Invariant (LTI) Systems

Objectives: After completing this chapter you should be able to do the following:

1.1. Describe (define) a linear system and apply this definition to test a given system for linearity.
1.2. Describe (define) a time-invariant system and apply this definition to test a given system for time-invariance.
1.3. Given one input-output pair, use the LTI properties to find the response to a second input, where the second input is simply related to the first input.

Rationale. What is the importance of LTI systems?

There are three methods for finding the response (output) of an LTI system: differential equations, transform methods, and convolution. These three methods are equivalent (under the conditions of system controllability and observability, which we will study later), and any one of the three may be used for an LTI system.

If the system is not linear then there is no general method for finding the output if the input is arbitrary. For some special cases a nonlinear differential equation that describes the system can be solved. Transform methods and convolution apply only to linear systems.

In short, we can easily solve problems for LTI systems, but not for nonlinear or time varying systems. Furthermore, the methods used depend on the LTI properties. Thus linearity and time invariance are of basic importance.

Objective 1.1. Describe (define) a linear system and apply this definition to test a given system for linearity.

In order to study linearity and time invariance of system models we will need some shorthand notation. Let us use the letter L to stand for the mathematical operation pictured in Fig. 1.1.1. That is, the letter L stands for whatever the system does to the input x in order to produce the output y. This is written as

$$y = L[x] \tag{1.1.1}$$

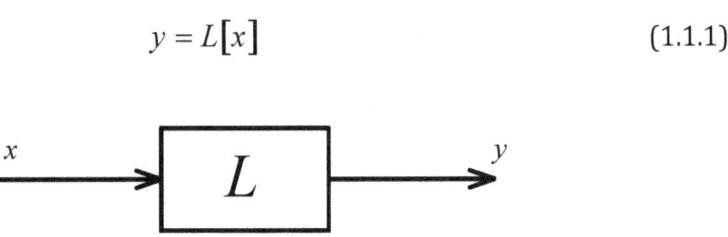

Fig. 1.1.1. System Model

Definition 1.1.1. Linearity. A system with an input-output relationship given by Eq. 1.1.1 is linear if

$$L[a_1 x_1(t) + a_2 x_2(t)] = a_1 L[x_1(t)] + a_2 L[x_2(t)] \tag{1.1.2}$$

for all $x_1, x_2, a_1, a_2,$

This definition really embodies two different criteria called additivity and homogeneity. If $a_1 = a_2 = 1$, then

$$L[x_1(t) + x_2(t)] = L[x_1(t)] + L[x_2(t)] \tag{1.1.3}$$

This property is called additivity. If $x_2(t) = 0$, then

$$L[a_1 x_1(t)] = a_1 L[x_1(t)] \tag{1.1.4}$$

This property is called homogeneity. Therefore a system is linear if it is both additive and homogeneous. Otherwise it is nonlinear.

Notes: a) The additivity and homogeneity properties are closely related. If $x_1 = x_2$ in Eq. 1.1.3 (additivity) then this is the same as $a = 2$ in Eq. 1.1.4 (homogeneity). In practical systems, additivity implies homogeneity, and therefore linearity. The converse is not true, however, as shown in Problem 1.1.2 below.

b) Superposition is another name for additivity. Therefore in practical systems superposition implies linearity.

c) The concept of linearity applies to much more than systems. In fact linearity permeates all of mathematics and the physical sciences.

d) There are no linear systems if we adhere strictly to definition 1.1.1. Any physical system will have limits such as saturation, maximum power, etc. It is our mathematical models that are linear.

The following are functional descriptions of linear system models, where $x(t)$ is the input and $y(t)$ is the output.

a) $\quad y(t) = 2x(t)$

b) $\quad y(t) = \int_{-\infty}^{t} x(\lambda)\,d\lambda$

c) $\quad y(t) = x(t)\cos(\omega t)$

Also, here are some functional descriptions of nonlinear system models.

d) $\quad y(t) = 2x(t) + 1$

e) $\quad y(t) = x^2(t)$

f) $\quad y(t) = x^2(t)\cos(\omega t)$

Problem 1.1.1. The current-voltage characteristic of a diode can often be approximated by $i(t) = v^2(t)$ for positive voltages as shown in Fig. 1.1.2. Determine whether or not this system is linear.

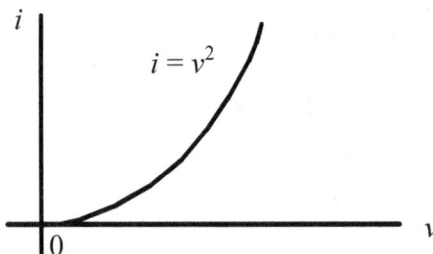

Fig. 1.1.2. Diode characteristic.

Solution: If $v(t)$ is the input and $i(t)$ the output then the operator of Eq. 1.1.1 is given by

$$i(t) = L[v(t)] = v^2(t), \quad v > 0$$

This system is nonlinear. It is neither additive nor homogeneous. To show this, first assume that $v(t) = v_1(t) + v_2(t)$, where both voltages are positive. Then

$$L[v(t)] = L[v_1(t) + v_2(t)]$$
$$= v_1^2(t) + 2v_1(t)v_2(t) + v_2^2(t)$$
$$= L[v_1(t)] + L[v_2(t)] + 2v_1(t)v_2(t)$$

The system is not additive because of the last term.

Next assume that $v(t) = av_1(t)$. Then

$$L[v(t)] = L[av_1(t)] = a^2 L[v_1(t)]$$

So the system is not homogeneous either.

Problem 1.1.2. Suppose a system consists of the two cascaded units shown in Fig. 1.1.3. The first sums $x^3 + \frac{d}{dt}x^3$, and the second unit takes the cube root of the sum. Determine whether or not this system is linear.

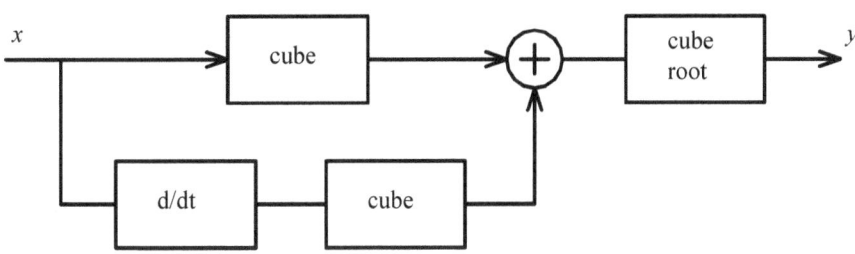

Fig. 1.1.3.

Solution: The response to x_1 is $y_1 = \sqrt[3]{x_1^3 + \frac{d}{dt}x^3}$. This system is homogeneous because if the input is multiplied by any constant a the output is given by

$$y = \sqrt[3]{(ax_1)^3 + \left(a\frac{d}{dt}x_1\right)^3} = a\sqrt[3]{x_1^3 + \frac{d}{dt}x_1^3} = ay_1$$

The system is not additive, however, for the response to $x_1 + x_2$ is given by

$$y = \sqrt[3]{(x_1 + x_2)^3 + \left(\frac{d}{dt}x_1 + \frac{d}{dt}x_2\right)^3} \neq y_1 + y_2$$

where y_1 is the response to x_1 and y_2 is the response to x_2. Therefore the system is not linear.

Note: This is an example of a system that is homogeneous but not additive. Thus a homogeneous system is not necessarily linear. It is true, however, that any practical additive system is also homogeneous.

Problem 1.1.3. Figure 1.1.4 shows a straight-line relationship, $y = x - 1$. Is this system linear?

Solution: Despite the fact that the relationship is a straight line, the system is not linear. Testing for additivity, suppose $x_1 = 2$ and $x_2 = 3$. Then

$$y_1 = x_1 - 1 = 1$$
$$y_2 = x_2 - 1 = 2$$

and the sum is 3. But the sum $x_1 + x_2$ is 5, and if this is the input the output is 4, not 3 as required by additivity.

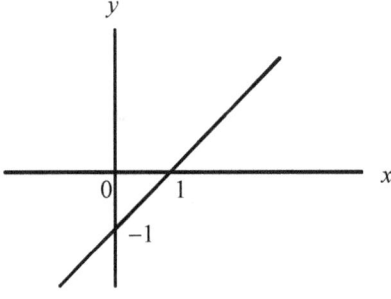

Fig. 1.1.4.

Notes: a) Neither is the system homogeneous (try it).
b) For the system to be linear, the straight line must pass through the origin.

Self Test, Objective 1.1.

a) Define a linear system.
b) With x the input and y the output, test the following systems for linearity.

a) $y = 2x + 3$ d) $y(t) = 3x^2(t)$

b) $y(t) = \int_{-\infty}^{t} x(\lambda)\,d\lambda$ e) $y(t) = x(t-1)$

c) $y(t) = 3x(t)$

(Self Test answers are given at the end of the chapter.)

Objective 1.2. Describe (define) a time-invariant system and apply this definition to test a given system for time-invariance.

Here is the definition of time-invariance.

Definition 1.2.1: Time-Invariance. A system with an input-output relationship given by Eq. 1.1.1 is time-invariant if

$$y(t+\varepsilon) = L[x(t+\varepsilon)] \quad \text{for all } \varepsilon \text{ and all } x \qquad (1.2.1)$$

Note: This implies that the response to the input is independent of the time the input is applied. If the input is delayed (or advanced) in time then the output is delayed (or advanced) by the same amount. Otherwise, there is no change.

Examples of time-invariant system models are:

a) $y(t) = 2x(t)$

b) $y(t) = x^2(t)$

c) $y(t) = \int_{-\infty}^{t} x(\lambda)\,d\lambda$

1.7

Also, here are some time-varying models.

d) $y(t) = x(t)\cos(\omega t)$

e) $y(t) = \int_0^t x(\lambda)\,d\lambda$

f) $y(t) = x^2\cos(\omega t)$

System (e) is time varying because of the lower limit on the integral. Look at Fig. 1.2.1. Since $y_2(t) \neq y_1(t+1/2)$ the system is time varying.

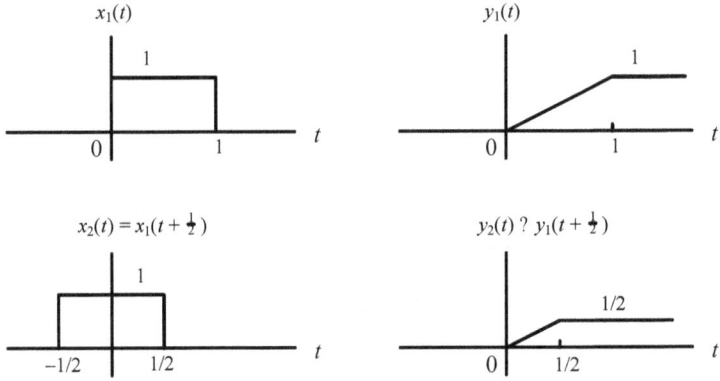

Fig. 1.2.1. Why (e) is time varying

Comparing these system models to those in Section 1.1.1 shows that a system may be classified into any one of four categories.

1. Linear time-invariant.
2. Linear time varying.
3. Nonlinear time-invariant.
4. Nonlinear time varying.

Of these, the first (abbreviated LTI) is the most important because we possess the mathematical tools to analyze these systems. Many practical examples fall into the other three categories. A select few of these can be analyzed.

Self Test. Objective 1.2.

a) Define a time-invariant system.

b) Test the following systems for time-invariance.

a) $y(t) = 3x(t)$ d) $y(t) = 3tx(t)$

b) $y(t) = 3x^2(t)$ e) $y(t) = x(t-1)$

c) $y(t) = \int_0^2 x(\lambda)\,d\lambda$

Objective 1.3. Given one input-output pair, use the LTI properties to find the response to a second input, where the second input is simply related to the first input.

The LTI constraint is exceedingly strong, for knowledge of a single input-output pair will (usually) provide enough information so that one can calculate the response to an arbitrary input. Read the following example, and then we will state what condition is necessary before a single input-output pair will completely characterize a system.

Look at Fig. 1.3.1. The response to $x_2(t)$, $x_3(t)$, and $x_4(t)$ can be calculated from the input-output pair $x_1(t)$, $y_1(t)$ by simply applying the LTI properties. Here's how:

Fig. 1.3.1b. Apply the homogeneity property. Since the input is doubled, the output is doubled.

1.9

Fig. 1.3.1c. Apply time-invariance and homogeneity. The input $x_3(t)$ is related to $x_1(t)$ by delay and multiplication by –1. Therefore the output $y_3(t)$ is found from $y_1(t)$ by delay and multiplication by –1.

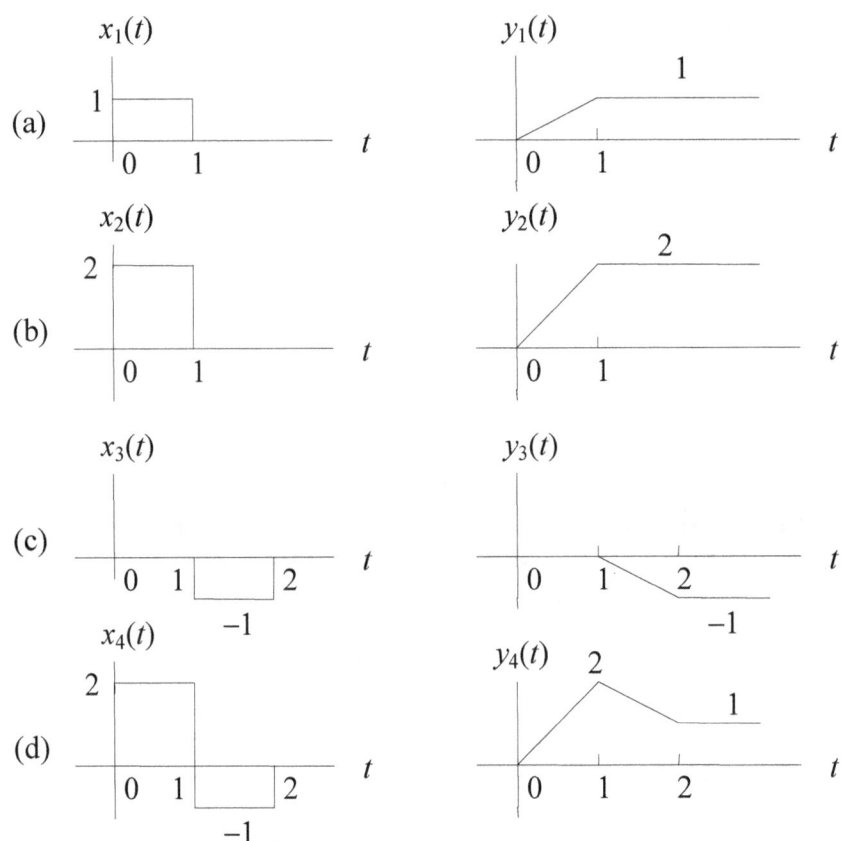

Fig. 1.3.1. Applying the LTI properties

Fig. 1.3.1d. Apply superposition. The input $x_4(t)$ is the sum $x_2(t) + x_3(t)$. Therefore the response $y_4(t)$ is the sum $y_2(t) + y_3(t)$.

In this example the input $x_4(t)$ is the sum of signals with known output. Therefore the response to $x_4(t)$ can be found from the $x_1(t)$, $y_1(t)$ pair by superposition. Thus knowledge of a single input-output pair, say $x_1(t), y_1(t)$, will allow us to find the response to an arbitrary signal, say $x_4(t)$, if we can express $x_4(t)$ as the sum of terms that are related to $x_1(t)$ in some elementary way.

In order to characterize an LTI system by an input-output pair, one should choose a pair that is "general" in the sense that most other input-output pairs can be determined from this one. This is possible. In fact, if we know the response to a unit step input, then the response to all other signals can be found. Since the unit impulse is linearly related to the unit step, knowledge of the unit impulse response allows us to find the response to arbitrary signals.

Some input-output pairs are not "general" enough. The response to a steady-state sinusoidal signal, for example, will be sufficient only for finding the response to any other sinusoid of the same frequency. But most input signals produce pairs that are general enough.

There are three closely related methods for finding the response of LTI systems to arbitrary input signals. These are convolution, differential or difference equations, and transform methods. In essence these three methods are nothing more than convenient ways to sum the responses to simple signals so as to find the response to a complicated signal, as for the $x_4(t), y_4(t)$ pair in Fig. 1.3.1. We will study each of these three methods.

Problem 1.3.1. An LTI system has the response $q(t)$ to the input $p(t)$ shown in Fig. 1.3.2.

a) Sketch the input $2p(t-1)$ and the corresponding response.

b) Express the input $x(t)$ shown in Fig. 1.3.3 as the sum

$$x(t) = \sum_n a_n p(t-n), \quad n = \cdots, -2, -1, 0, 1, 2, 3, \cdots$$

That is, find each value of a_n.

c) Find and sketch $y(t)$, the response to $x(t)$.

1.11

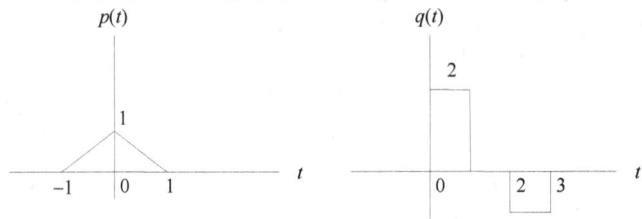

Fig. 1.3.2. An input-output pair.

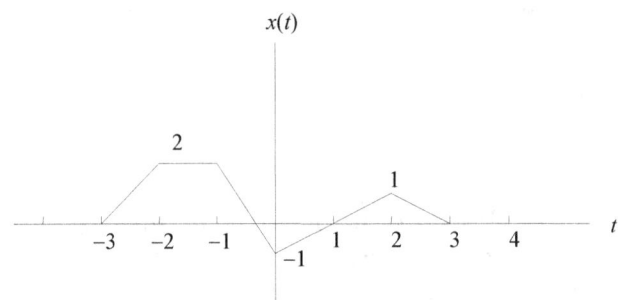

Fig. 1.3.3. The input x(t)

Solution: a) Figure 1.3.4 shows $2p(t-1)$ and the response $2q(t-1)$.

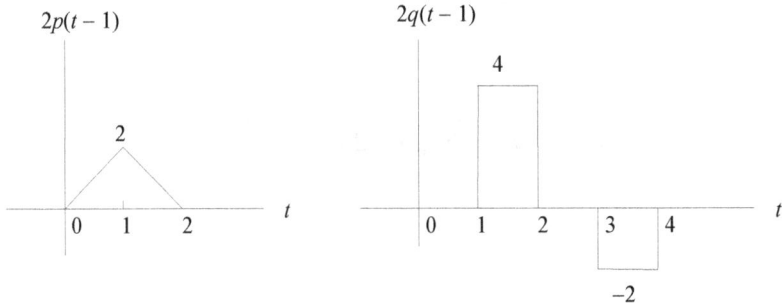

Fig. 1.3.4.

1.12

b) Figure 1.3.5 shows the various versions of $p(t)$ that combine to make up $x(t)$, giving

$$x(t) = 2p(t+2) + 2p(t+1) - p(t) + p(t-2)$$

Thus, $a_{-2} = 2$, $a_{-1} = 2$, $a_0 = -1$, $a_1 = 0$, $a_2 = 1$. All others are zero.

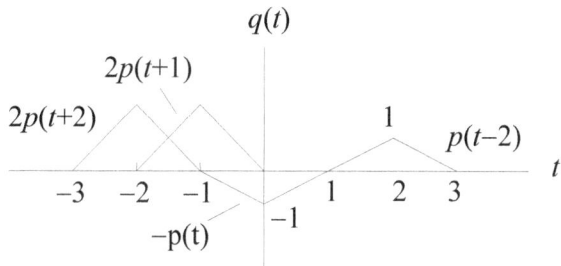

Fig. 1.3.5.

c) The output $y(t)$ in Fig. 1.3.6 is given by

$$y(t) = \sum_n a_n q(t-n)$$
$$= 2q(t+2) + 2q(t+1) - q(t) + q(t-2)$$

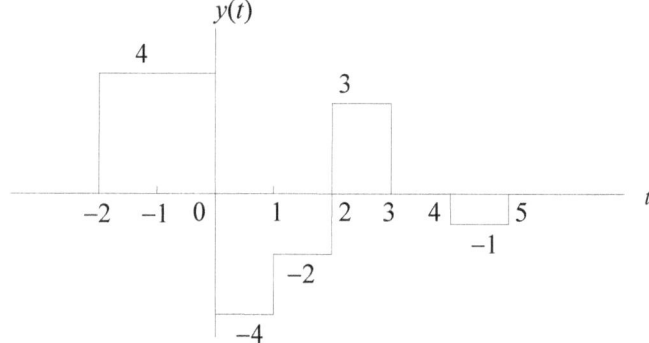

Fig. 1.3.6. The output $y(t)$

1.13

Problem 1.3.2. In Fig. 1.3.7 the input-output pair for an LTI system is $x_1(t), y_1(t)$. Find the response to $x_2(t)$.

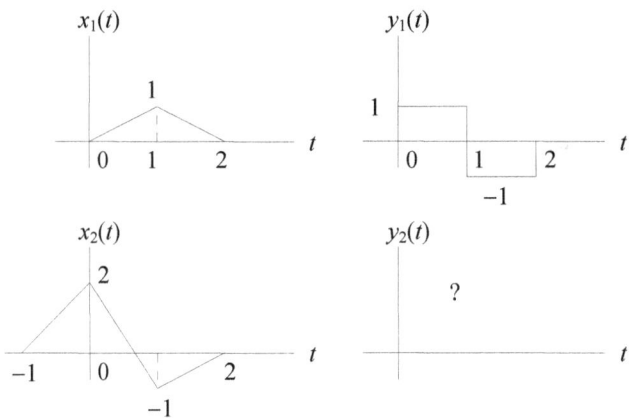

Fig. 1.3.7.

Figure 1.3.8 shows the answer.

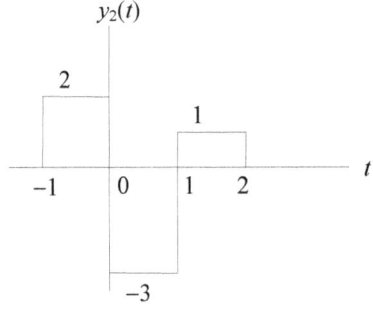

Fig. 1.3.8.

1.14

Problem 1.3.3. The output of an LTI system is $q(t)$ if the input is $p(t)$ shown in Fig. 1.3.9. Find the response to $x(t)$.

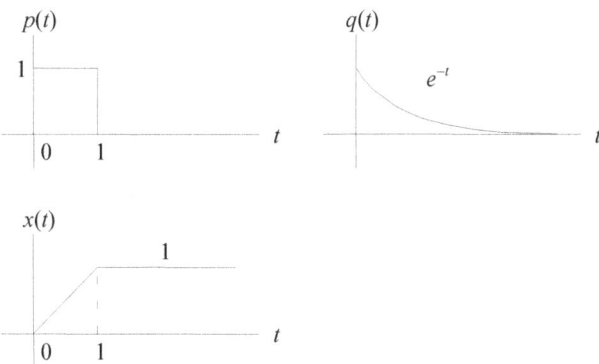

Fig. 1.3.9.

Solution: The method used in Problems 1.3.1 and 1.3.2 will not work here because $x(t)$ cannot be expressed as the sum of signals simply related to $p(t)$. But

notice that $x(t)$ is the integral of $p(t)$, $x(t) = \int_{-\infty}^{t} p(\lambda)\,d\lambda$. Integration is summation.

Instead of summing discrete items, integration sums continuous items. This brings superposition into play, making $y(t)$ the integral of $q(t)$.

$$y(t) = \int_{-\infty}^{t} q(\lambda)\,d\lambda = 1 - e^{-t}, \quad t > 0$$

as shown in Fig. 1.3.10. To put it another way, if $x(t)$ is the integral of $p(t)$ then $y(t)$ must be the integral of $q(t)$.

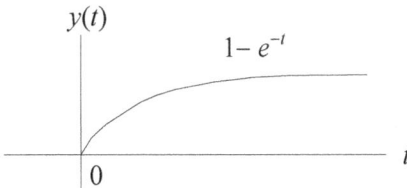

Fig. 1.3.10

1.15

Notes: a) Integration and differentiation are linear operations. It therefore follows that for an LTI system, integration (differentiation) of the input produces an output that is the integral (derivative) of the original output. That is, if x_1, y_1 is an input-output pair, then x_2, y_2 form an input-output pair, where

$$x_2(t) = \int_{-\infty}^{t} x_1(\lambda)\,d\lambda \qquad\qquad y_2(t) = \int_{-\infty}^{t} y_1(\lambda)\,d\lambda$$

also, $x_3(t) = \frac{d}{dt}x_1(t)$ and $y_3(t) = \frac{d}{dt}y_1(t)$ form an input-output pair.

b) In Problem 1.3.2 $y_1(t)$ is the derivative of $x_1(t)$. Notice that $y_2(t)$ is the derivative of $x_2(t)$.

Problem 1.3.4. When a unit step voltage $x_1(t) = u(t)$ is applied to an LTI system the response is $\qquad y_1(t) = \left(\frac{1}{2} - \frac{1}{2}e^{-10t}\right)u(t)$.

a) Find the response to a unit ramp, $x_2(t) = tu(t)$.

b) Find the response to a unit impulse, $x_3(t) = \delta(t)$.

Solution: a) Since $x_2(t)$ is the integral of $x_1(t)$ then $y_2(t)$ is the integral of $y_1(t)$.

$$y_2(t) = \int_{-\infty}^{t} y_1(\lambda)\,d\lambda = \int_{0}^{t}\left(\frac{1}{2} - \frac{1}{2}e^{-10\lambda}\right)d\lambda$$

$$= \frac{1}{2}\lambda + \frac{1}{20}e^{-10\lambda}\Big|_{0}^{t} = \left[\frac{t}{2} - \frac{1}{20}\left(1 - e^{-10t}\right)\right]u(t)$$

b) Since $x_3(t)$ is the derivative of $x_1(t)$, differentiate $y_1(t)$ to obtain

$$y_3(t) = \frac{d}{dt}y_1(t) = 5e^{-10t}u(t)$$

1.16

Self Test, Objective 1.3.

Functions $p(t)$, $q(t)$ in Fig. 1.3.11 form an input-output pair for an LTI system. Find the response to $x(t)$.

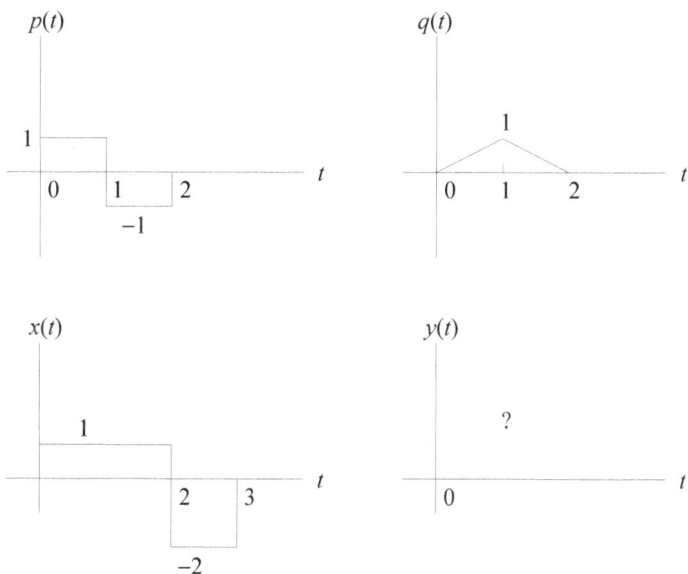

Fig. 1.3.11

1.17

Self Test Answers:

Objective 1.1.

a) See Definition 1.1.1.

b) Systems b, c, and e are linear.

Objective 1.2.

a) See Definition 1.2.1.

b) Systems a, b, and e are time-invariant.

Objective 1.3.

Since $x(t) = p(t) + 2p(t - 1)$ then $y(t) = q(t) + 2q(t - 1)$ as shown in Fig. 1.3.12.

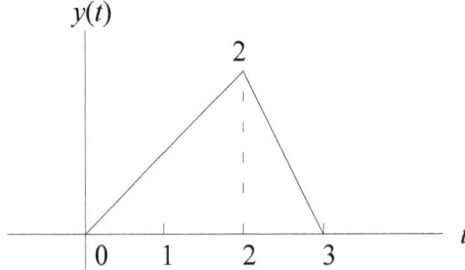

Fig. 1.3.12.

Chapter 2.

Formulation Procedures

How to derive the state model for a circuit.

Objective: After completing this chapter you should be able to do the following:

2.1. Write the state equations for a given electric circuit.

Rationale. Why learn to write state equations?

We begin our study of the first of three methods for finding the response of an LTI system, namely differential equations. This chapter investigates how to write differential equations in state variable form. Then Chapters 3 and 4 present methods of solution.

Differential equations that describe systems can be in one of two forms, a single n^{th} order equation or n first order equations. This second form is the state formulation. The state variable model is in many ways more natural than the single n^{th} order equation. It can be written in a more compact form using matrix notation and it lends itself to approximate numerical solution, as we will see in Chapter 5.

Chapter 2 Pre-Test:

Before studying this chapter you should be able to do the following:

1. Re-write the following three equations in matrix form.

$$y_1 = 3x_1 - 2x_2 + x_3 + 2$$

$$y_2 = x_1 + 2x_2 - x_3$$

$$y_3 = 2x_1 - 4x_2 + 3x_3 - 3$$

2. Re-write the following matrix equation as three separate equations.

$$\begin{bmatrix} y_1 \\ y_2 \\ y_3 \end{bmatrix} = \begin{bmatrix} 2 & -2 & 6 \\ 1 & 0 & 3 \\ 4 & -2 & 0 \end{bmatrix} \begin{bmatrix} x_1 \\ x_2 \\ x_3 \end{bmatrix} + \begin{bmatrix} 2 \\ -2 \\ 0 \end{bmatrix}$$

(Answer given at the end of chapter.)

Objective 2.1. Write the state equations for a given electric circuit.

2.1.1. Introduction

This chapter introduces procedures for formulating the state equations of networks in canonical form. For networks containing no tie sets (loops) of capacitors and/or voltage sources, or cut sets (nodes) of inductors and/or current sources, this form is given by

$$\dot{x} = Ax + Bq \tag{2.1.1}$$

Here, x is the state variable, q is the forcing function, A and B are constant matrices, and the dot above x denotes derivative. In longhand notation Eq. 2.1.1 is written as

$$\frac{dx_1}{dt} = a_{11}x_1 + a_{12}x_2 + \cdots + a_{1n}x_n + b_{11}q_1 + \cdots + b_{1r}q_r$$

$$\frac{dx_2}{dt} = a_{21}x_1 + a_{22}x_2 + \cdots + a_{2n}x_n + b_{21}q_1 + \cdots + b_{2r}q_r$$

$$\vdots \qquad\qquad \vdots$$

$$\frac{dx_n}{dt} = a_{n1}x_1 + a_{n2}x_2 + \cdots + a_{nn}x_n + b_{n1}q_1 + \cdots + b_{nr}q_r$$

The term canonical form means that each equation contains the first derivative of only one state variable.

For electrical networks each state variable x_1, x_2, ..., x_n is a voltage or current in the network, and each forcing function q_1, q_2, ..., q_r is either a voltage source or current source. If a network does contain capacitor-voltage source tie sets or inductor-current source cut sets, then the canonical form is given by

$$\dot{x} = Ax + Bq + E\dot{q} \tag{2.1.2}$$

That is, the forcing function may contain the first derivative of the current sources or the voltage sources.

Equation 2.1.2 is the most general form of the state equations, and it is a complete mathematical description of the network if we include the initial values of the state variables, $x_1(t_0), x_2(t_0), ..., x_n(t_0)$. "Complete mathematical description" means that any network variable (any current or voltage) can be expressed as a linear combination of the state variables, the sources, and the first derivatives of the sources.

2.1.2 Type A and B Networks

First let's describe cut sets and tie sets. A cut set can be defined in two equivalent ways. A set of branches that intersect a closed surface is called a cut set of the network. In Fig. 2.1.1 the closed surface C intersects branches 1, 2, 3, 5, 6. These branches form a cut set. Note that Kirchhoff's current law must be satisfied by the branch currents of any cut set. Thus we can write $0 = i_1 - i_2 - i_3 - i_5 - i_6$. The closed surface C is called a node, or in the case that it encloses one or more elements as in Fig. 2.1.1, it is called a super node.

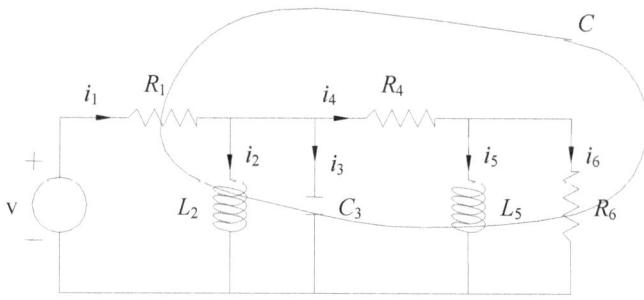

Fig. 2.1.1. Illustrating a cut set.

An alternate definition for cut set is a set of branches that satisfy both of the following conditions.

1. If all these branches are cut the network is in two separate parts.

2. If all but one of these branches are cut the network is not in two separate parts.

Now look at Fig. 2.1.2. This is the same network shown in Fig. 2.1.1. The branches 3, 5, and 6 intersect the closed curve C but they do not form a cut set. Neither of our above definitions is satisfied, and as a further check the sum $i_3 + i_5 + i_6$ is not required to equal zero by Kirchhoff's current law. If you use the first definition, the closed curve C must pass through elements (branches) and not through nodes as it does between R_1 and R_4 in Fig. 2.1.2.

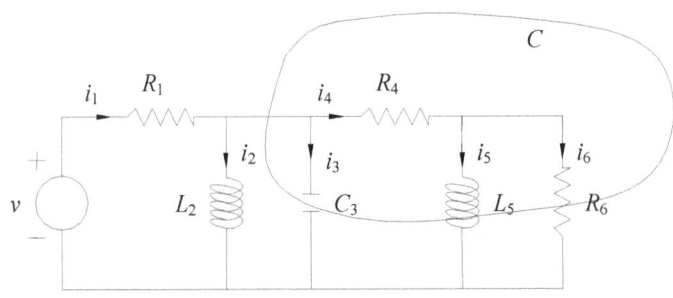

Fig. 2.1.2. Not a cut set

The set of branches in a loop is called a tie set. In Fig. 2.1.2 the voltage source v along with R_1, R_4, and L_5 form a tie set. Note that Kirchhoff's voltage law must be satisfied around any tie set.

Definition 2.1.1 Type A and B networks. Type A networks contain no tie sets of capacitors and/or voltage sources, or no cut sets of inductors and/or current sources. Type B networks contain one or both of these.

The network in Fig. 2.1.2 is a type A network. Two examples of type B networks are shown in Fig. 2.1.3. In Fig. 2.1.3a there is a tie set consisting of C_1,

C_4, and the voltage source. In Fig. 2.1.3b there is a cut set consisting of L_2, L_3, and the current source.

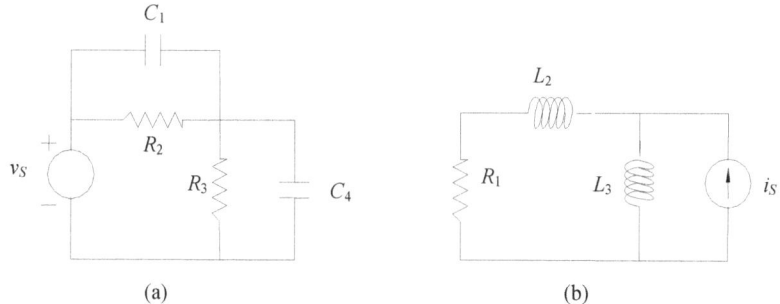

(a)   (b)

Fig. 2.1.3. Type B networks

We will first describe the procedure for deriving type A equations (Eq. 2.1.1). Then this procedure will be modified for type B networks (Eq. 2.1.2.) Before proceeding, answer the following questions:

1. What does "the canonical form of state equations" mean?
2. Write the general form for type A state equations.
3. Write the general form for type B state equations.
4. Describe (define) type A and B networks.
5. Describe (define) a cut set.
6. Describe (define) a tie set.

2.1.3 Formulation Procedures – Type A Networks

In this procedure the inductor currents and capacitor voltages will be the state variables. Our objective is to wind up with equations containing only the state variables and source variables. So the first step is to identify these important variables.

<u>Step 1.</u> Write down the state variables and source variables. Using Fig. 2.1.4 to illustrate gives

CHAPTER 2

$\boxed{i_3, v_4, i_5, v_6}$

Step 2. Write down the inductor and capacitor component equations.

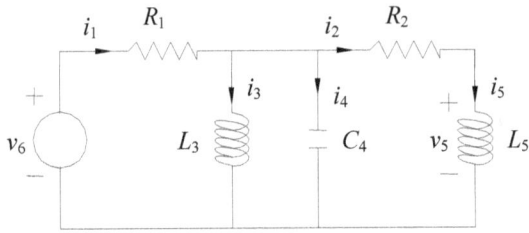

Fig. 2.1.4. Example Network

$$L_3 \frac{di_3}{dt} = v_3 \qquad (2.1.3)$$

$$C_4 \frac{dv_4}{dt} = i_4 \qquad (2.1.4)$$

$$L_5 \frac{di_5}{dt} = v_5 \qquad (2.1.5)$$

Step 3. The remainder of the procedure is simple enough if you know where you are going. And we're going to express the variables on the right side of these equations in terms of the variables in Step 1 (shown in the box.) Use the tie set and cut set along with the remaining component equations to do this.

From Fig. 2.1.4 note that $v_3 = v_4$. Therefore rewrite Eq. 2.1.3 as

$$L_3 \frac{di_3}{dt} = v_4$$

All variables in this equation are in the box, so this is the first state equation.

Next, let's work on i_4 in Eq. 2.1.4. Write the cut set equation given by

$$i_1 = i_3 + i_4 + i_5$$

or
$$i_4 = i_1 - i_3 - i_5$$

Variables i_3 and i_5 are in the box. To express i_1 in terms of box variables, write

2.6

$$i_1 = v_1 / R_1$$

Since
$$v_1 = v_6 - v_4$$

This gives
$$i_4 = \frac{v_6 - v_4}{R_1} - i_3 - i_5$$

Substituting this into Eq. 2.1.4 gives our second state equation.

$$C_4 \frac{dv_4}{dt} = \frac{v_6 - v_4}{R_1} - i_3 - i_5$$

Now to tackle the third state equation we need to express v_5 in terms of the variables in the box. The rightmost loop gives

$$v_5 = v_4 - v_2$$

Since $i_2 = i_5$, express v_2 as $R_2 i_5$. Therefore

$$v_5 = v_4 - R_2 i_5$$

Substitute into Eq. 2.1.5 to obtain the third state equation.

$$L_5 \frac{di_5}{dt} = v_4 - R_2 i_5$$

After dividing by constants L_3, C_4, and L_5, these equations can be displayed in matrix form as

$$\begin{bmatrix} \dot{i}_3 \\ \dot{v}_4 \\ \dot{i}_5 \end{bmatrix} = \begin{bmatrix} 0 & \frac{1}{L_3} & 0 \\ \frac{-1}{C_4} & \frac{-1}{R_1 C_4} & \frac{-1}{C_4} \\ 0 & \frac{1}{L_5} & \frac{-R_2}{L_5} \end{bmatrix} \begin{bmatrix} i_3 \\ v_4 \\ i_5 \end{bmatrix} + \begin{bmatrix} 0 \\ \frac{1}{R_1 C_4} \\ 0 \end{bmatrix} v_6$$

Problem 2.1.1. Write the state equations for the series RLC circuit in Fig. 2.1.5.

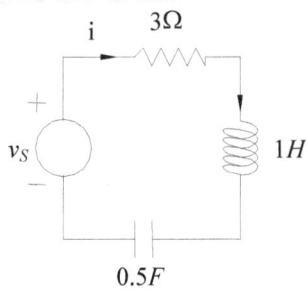

Fig. 2.1.5. A series RLC circuit

Solution: There are two energy storage elements, so there are two state variables, the inductor current and the capacitor voltage.

<u>Step 1.</u> List the source and state variables $\left| v_S, i_L, v_C \right|$.

<u>Step 2.</u> Write the inductor and capacitor component equations.

$$\frac{di_L}{dt} = v_L$$

$$\frac{1}{2}\frac{dv_C}{dt} = i_C$$

Step 3. Express v_L in terms of the box variables from the tie set and resistor component equation to get

$$v_L = v_S - 3i_L - v_C$$

Since $i_C = i_L$, this gives the state equations:

$$\frac{di_L}{dt} = v_S - 3i_L - v_C$$

$$\frac{dv_C}{dt} = 2i_L$$

Problem 2.1.2. Write the state equations for the parallel RLC circuit driven by a current source as shown in Fig. 2.1.6.

2.8

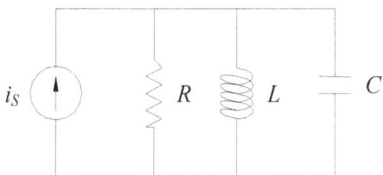

Fig. 2.1.6. A parallel RLC circuit.

Answer:

$$L\frac{di_L}{dt} = v_C$$

$$C\frac{dv_C}{dt} = -\frac{v_C}{R} - i_L + i_S$$

Problem 2.1.3. So far we have not encountered a circuit containing more than one independent source. Write the state equations for the circuit in Fig. 2.1.7.

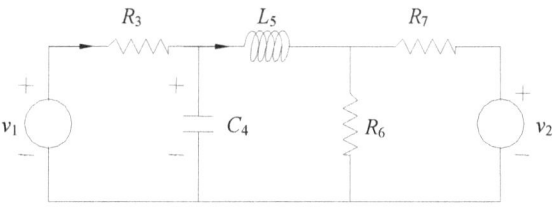

Fig. 2.1.7.

Answer:

$$C_4\frac{dv_4}{dt} = -\frac{v_4}{R_3} - i_5 + \frac{v_1}{R_3}$$

$$L_5\frac{di_5}{dt} = v_4 - \frac{R_6 R_7}{R_6 + R_7}i_5 - \frac{R_6}{R_6 + R_7}v_2$$

(Hint: Find Thevenin's equivalent circuit for the R_6, R_7, v_2 network.)

2.9

2.1.4. Formulation Procedures – Type B Networks.

Type B networks contain a tie set of capacitors and/or voltage sources, or a cut set of inductors and/or current sources, so the state equations will be of the form or Eq. 2.1.2. That is, the forcing function will contain the derivative of any source in such a tie set or cut set.

To get an inkling of why this is so, consider Fig. 2.1.8. Writing the capacitor component equations in an attempt to formulate the state equations gives

$$C_1 \frac{dv_1}{dt} = i_1 \tag{2.1.6}$$

$$C_4 \frac{dv_4}{dt} = i_4 \tag{2.1.7}$$

But since $v_4 = v_S - v_1$, the current i_4 in Eq. 2.1.7 can be expressed in terms of the known source voltage v_S and capacitor voltage v_1 as

$$C_4 \frac{dv_S}{dt} - C_4 \frac{dv_1}{dt} = i_4 \tag{2.1.8}$$

Note two things: The derivative of the source voltage appears in this equation, and there is only one state variable v_1 between Eqs. 2.1.6 and 2.1.7.

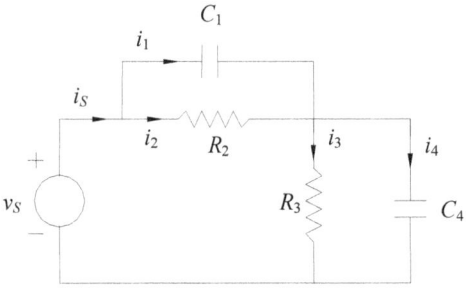

Fig. 2.1.8. Type B Network

Here is the procedure for Type B networks:

<u>Step 1.</u> List the state and source variables in a box. For a tie set of capacitors and/or voltage sources include all but one of the capacitor voltages. For a cut set

of inductors and/or current sources include all but one of the inductor currents. For the circuit in Fig. 2.1.8 the box contains

$$\left| v_1, v_S \right|$$

Step 2. Write the component equations for all capacitors and inductors. (Eqs. 2.1.6 and 2.1.7.)

Step 3. Eliminate the extra equation by using the cut set or tie set relation. (Eq.. 2.1.8.)

Step 4. Express the variables on the right side of the remaining equations in terms of the variables in the box. Use the tie set and cut set along with the remaining component equations to do this.

To continue our derivation of the state equation in Fig. 2.1.8, write the following cut set equation.

$$i_4 = i_1 + i_2 - i_3$$

Write component equations to find i_1, i_2, and i_3.

$$i_1 = C_1 \frac{dv_1}{dt}$$

$$i_2 = \frac{v_1}{R_2}$$

$$i_3 = \frac{v_3}{R_3} = \frac{v_S - v_1}{R_3}$$

Substitute into Eq. 2.1.8 to obtain the state equation.

$$C_4 \frac{dv_5}{dt} - C_4 \frac{dv_1}{dt} = C_1 \frac{dv_1}{dt} + \frac{v_1}{R_2} - \frac{v_S - v_1}{R_3}$$

Problem 2.1.4. Write the state equation for the circuit in Fig. 2.1.9.

2.11

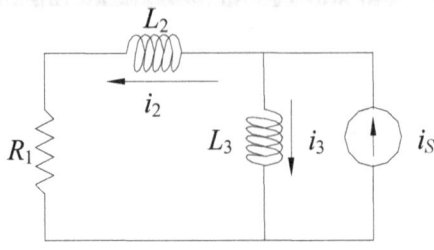

Fig. 2.1.9.

Solution: For this type B network choose i_2 as the state variable. The box consists of

$$|\,i_2,\,i_s\,|$$

The component equation for L_2 is given by

$$L_2 \frac{di_2}{dt} = v_2$$

Now express v_2 in terms of the variables in the box. The tie set equation gives

$$v_2 = v_3 - v_1$$

where

$$v_3 = L_3 \frac{di_3}{dt} = L_3 \frac{d(i_3 - 1_2)}{dt}$$

and

$$v_1 - i_2 R_1$$

so the state equation is given by

$$(L_2 + L_3) \frac{di_2}{dt} = -R_1 i_2 + L_3 \frac{di_s}{dt}$$

Problem 2.1.5. Write the state equations for the circuit in Fig. 2.1.10. Use v_3 and i_4 as the state variables.

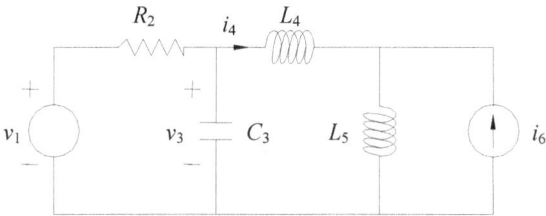

Fig. 2.1.10.

Answer:

$$\frac{dv_3}{dt} = -\frac{1}{C_3 R_2} v_3 - \frac{1}{C_3} i_4 + \frac{1}{C_3 R_2} v_1$$

$$\frac{di_4}{dt} = \frac{1}{L_4 + L_5} v_3 - \frac{L_5}{L_4 + L_5} \frac{di_6}{dt}$$

Problem 2.1.6. Write the state equations for the circuit in Fig. 2.1.11. Use v_2 and i_5 as the state variables.

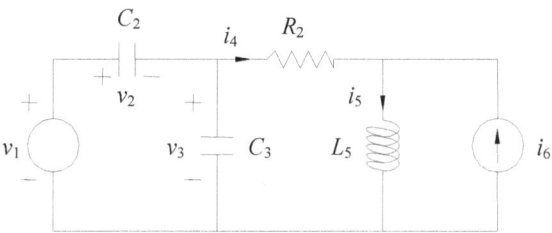

Fig. 2.1.11.

Answer:

$$(C_2 + C_3) \frac{dv_2}{dt} = i_5 - i_6 + C_3 \frac{dv_1}{dt}$$

$$L_5 \frac{di_5}{dt} = v_1 - R_4 i_5 - v_2 + R_4 i_6$$

2.13

Self Test, Objective 2.1.

Write the state equations in matrix form for the networks in Figs. 2.1.12 and 2.1.13.

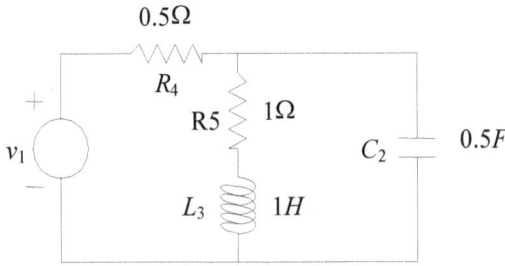

Fig. 2.1.12.

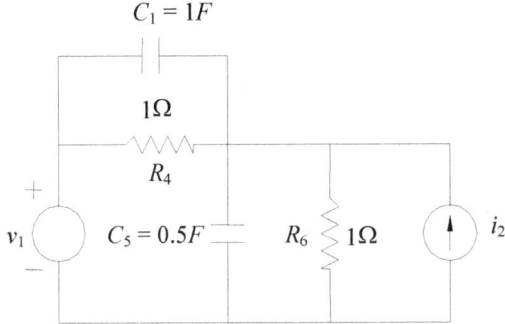

Fig. 2.1.13.

Pre-Test Answers:

1. $$\begin{bmatrix} y_1 \\ y_2 \\ y_3 \end{bmatrix} = \begin{bmatrix} 3 & -2 & 1 \\ 1 & 2 & -1 \\ 2 & -4 & 3 \end{bmatrix} \begin{bmatrix} x_1 \\ x_2 \\ x_3 \end{bmatrix} + \begin{bmatrix} 2 \\ 0 \\ 3 \end{bmatrix}$$

2. $$y_1 = 2x_1 - 2x_2 + 6x_3 + 2$$

$$y_2 = x_1 + 3x_3 - 2$$

$$y_3 = 4x_1 - 2x_2$$

Self Test Answers:

First for Fig. 2.1.12.

$$\begin{bmatrix} \dot{i}_3 \\ \dot{v}_2 \end{bmatrix} = \begin{bmatrix} -1 & 1 \\ -2 & -4 \end{bmatrix} \begin{bmatrix} i_3 \\ v_2 \end{bmatrix} + \begin{bmatrix} 0 \\ 4 \end{bmatrix} v_1$$

or

$$\begin{bmatrix} \dot{v}_2 \\ \dot{i}_3 \end{bmatrix} = \begin{bmatrix} -4 & -2 \\ 1 & -1 \end{bmatrix} \begin{bmatrix} v_2 \\ i_3 \end{bmatrix} + \begin{bmatrix} 4 \\ o \end{bmatrix} v_1$$

For Fig. 2.1.13, either

$$\dot{v}_3 = -\tfrac{4}{3} v_3 + \tfrac{2}{3} v_1 - \tfrac{2}{3} i_2 + \tfrac{1}{3} \dot{v}_1$$

or

$$\dot{v}_5 = -\tfrac{4}{3} v_5 + \tfrac{2}{3} v_1 + \tfrac{2}{3} i_2 + \tfrac{2}{3} \dot{v}_1$$

Chapter 3
The Solution of State Equations -Type A Networks

Objectives: After completing this chapter you should be able to do the following:

3.1. Find a closed form solution of the state equations for type A networks.

Rationale: Both a closed form solution as well as a series solution of the LTI state equations can be found. The closed form analytical solution can be used to predict system performance – if our state formulation is a valid model of the system. Obviously this is a useful thing to know for a practicing electrical engineer.

In this and the next chapter we study the closed form solution. The series solution is useful for approximation purposes and will be used in Chapter 5.

Pre-Test:

1. Solve for x_1 and x_2.

$$\begin{bmatrix} 2 \\ -1 \end{bmatrix} = \begin{bmatrix} 1 & -1 \\ 3 & 1 \end{bmatrix}\begin{bmatrix} x_1 \\ x_2 \end{bmatrix}$$

2. If A is given by

$$\begin{bmatrix} 1 & -1 \\ 3 & 1 \end{bmatrix}$$

find the value of the determinant $|2I - A|$ where I is the identity matrix.

CHAPTER 3

Objective 3.1: Find a closed form solution of the state equations for type A networks.

3.1.1. First Order Equations – A Review.

The state equations for a linear network are first-order matrix equations. There are several methods for solving linear first order equations. Since most readers are familiar with the method of undetermined coefficients, this chapter extends this method from simple first-order equations to matrix equations. Let us begin with a review of this technique as it applies to simple first-order equations.

For concreteness, consider the equation

$$\frac{dv}{dt} = -2v + 2e^{-t}, \quad v(0) = 0 \tag{3.1.1}$$

The last term is the forcing function, and the initial value of v is given. Since the equation is linear, the homogeneous (source-free) response and the particular (forced) response can be found separately. Then add the two solutions and use the initial conditions to solve for the arbitrary constants. This same procedure applies to matrix equations.

The Particular Solution

The particular solution to Eq. 3.1.1 is proportional to the forcing function plus all its derivatives. Since all derivatives are of the form of the forcing function itself, choose

$$v_p(t) = Ke^{-t}$$

where K is a constant. Substituting this function into Eq. 3.1.1 gives

$$-Ke^{-t} = -2Ke^{-t} + 2e^{-t}$$

or

$$-K = -2K + 2$$

which gives $K = 2$. Therefore

$$v_p(t) = 2e^{-t} \tag{3.1.2}$$

The Homogeneous solution

Set the forcing function to zero to obtain the homogeneous equation. Equation 3.1.1 becomes

$$\frac{dv_h}{dt} = -2v_h \qquad (3.1.3)$$

Assume an exponential solution given by

$$v_h(t) = ae^{st}$$

Substitute this into Eq. 3.1.3 to get the characteristic equation, given by

$$s = -2$$

Therefore the homogeneous solution is given by

$$v_h(t) = ae^{-2t} \qquad (3.1.4)$$

Find the complete solution by adding Eqs. 3.1.2 and 3.1.4 to obtain

$$v(t) = 2e^{-t} + ae^{-2t}$$

Now solve for the constant a from the initial condition to finally obtain

$$v(t) = 2e^{-t} - 2e^{-2t}, \quad t > 0 \qquad (3.1.5)$$

3.1.2. Closed Form Solution of the State Equations.

The general features of the solution of state equations can be illustrated by considering a second order network. As an example, consider the circuit in Fig. 3.1.1.

The state equations for this network were found in Problem 2.1.1.

$$\begin{bmatrix} \dot{v} \\ \dot{i} \end{bmatrix} = \begin{bmatrix} 0 & 2 \\ -1 & -3 \end{bmatrix} \begin{bmatrix} v \\ i \end{bmatrix} + \begin{bmatrix} 0 \\ 1 \end{bmatrix} e^{-1.5t} \qquad (3.1.6)$$

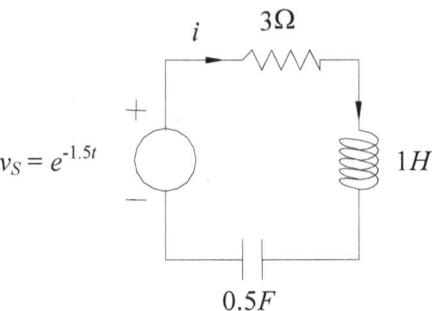

Fig. 3.1.1. A Second-Order Circuit

Assume the network to be initially at rest with the source voltage connected at $t = 0$. Hence the initial conditions are

$$\begin{bmatrix} v(0) \\ i(0) \end{bmatrix} = \begin{bmatrix} 0 \\ 0 \end{bmatrix}$$

The Particular Solution

The particular solution to Eq. 3.1.6 is proportional to the forcing function plus all its derivatives. Therefore choose

$$\begin{bmatrix} v_p \\ i_p \end{bmatrix} = \begin{bmatrix} K_1 \\ K_2 \end{bmatrix} e^{-1.5t}$$

Substitute this into Eq. 3.1.6 to obtain

$$\begin{bmatrix} -1.5K_1 \\ -1.5K_2 \end{bmatrix} = \begin{bmatrix} 0 & 2 \\ -1 & -3 \end{bmatrix} \begin{bmatrix} K_1 \\ K_2 \end{bmatrix} + \begin{bmatrix} 0 \\ 1 \end{bmatrix}$$

from which $K_1 = -8$, $K_2 = 6$. Therefore the particular solution is given by

$$\begin{bmatrix} v_p \\ i_p \end{bmatrix} = \begin{bmatrix} -8 \\ 6 \end{bmatrix} e^{-1.5t} \tag{3.1.7}$$

The Homogeneous Solution

Setting the forcing function to zero in Eq. 3.1.6 gives

$$\begin{bmatrix} \dot{v}_h \\ \dot{i}_h \end{bmatrix} = \begin{bmatrix} 0 & 2 \\ -1 & -3 \end{bmatrix} \begin{bmatrix} v_h \\ i_h \end{bmatrix} \tag{3.1.8}$$

Assume an exponential solution given by

$$\begin{bmatrix} v_h \\ i_h \end{bmatrix} = \begin{bmatrix} a \\ b \end{bmatrix} e^{st}$$

Then the derivative vector is

$$\begin{bmatrix} \dot{v}_h \\ \dot{i}_h \end{bmatrix} = \begin{bmatrix} sa \\ sb \end{bmatrix} e^{st}$$

Substituting into Eq. 3.1.8 gives

$$\begin{bmatrix} sa \\ sb \end{bmatrix} = \begin{bmatrix} 0 & 2 \\ -1 & -3 \end{bmatrix} \begin{bmatrix} a \\ b \end{bmatrix}$$

A more convenient form is

$$\begin{bmatrix} s & 0 \\ 0 & s \end{bmatrix} \begin{bmatrix} a \\ b \end{bmatrix} = \begin{bmatrix} 0 & 2 \\ -1 & -3 \end{bmatrix} \begin{bmatrix} a \\ b \end{bmatrix}$$

giving

$$\begin{bmatrix} s & -2 \\ 1 & (s+3) \end{bmatrix} \begin{bmatrix} a \\ b \end{bmatrix} = \begin{bmatrix} 0 \\ 0 \end{bmatrix} \tag{3.1.9}$$

Since Eq. 3.1.9 must hold for arbitrary constants a and b, the determinant of the coefficient matrix must be zero. That is,

$$\begin{vmatrix} s & -2 \\ 1 & (s+3) \end{vmatrix} = s^2 + 3s + 2 = 0 \tag{3.1.10}$$

This is the characteristic equation. Solving for s gives the two values –1, –2. Thus there are two homogeneous solutions, each with arbitrary constants, given by

$$\begin{bmatrix} v_h \\ i_h \end{bmatrix} = \begin{bmatrix} a_1 \\ b_1 \end{bmatrix} e^{-t} + \begin{bmatrix} a_2 \\ b_2 \end{bmatrix} e^{-2t} \qquad (3.1.11)$$

Now add the particular and homogeneous solutions to obtain the complete solution given by

$$\begin{bmatrix} v \\ i \end{bmatrix} = \begin{bmatrix} -8 \\ 6 \end{bmatrix} e^{-1.5t} + \begin{bmatrix} a_1 \\ b_1 \end{bmatrix} e^{-t} + \begin{bmatrix} a_2 \\ b_2 \end{bmatrix} e^{-2t} \qquad (3.1.12)$$

Solve for the four constants a_1, a_2, b_1, b_2. This requires four linear independent equations. Two equations can be obtained directly from Eq. 3.1.12 by substituting the initial conditions.

$$\begin{bmatrix} 0 \\ 0 \end{bmatrix} = \begin{bmatrix} -8 \\ 6 \end{bmatrix} + \begin{bmatrix} a_1 \\ b_1 \end{bmatrix} + \begin{bmatrix} a_2 \\ b_2 \end{bmatrix}$$

or in longhand notation these equations are given by

$$a_1 + a_2 = 8$$

$$b_1 + b_2 = -6$$

Use Eq. 3.1.9 for the other two equations. For $s = -1$ we have

$$\begin{bmatrix} -1 & -2 \\ 1 & 2 \end{bmatrix} \begin{bmatrix} a_1 \\ b_1 \end{bmatrix} = \begin{bmatrix} 0 \\ 0 \end{bmatrix}$$

or two identical equations, given by

$$a_1 + 2b_1 = 0$$

Now substitute $s = -2$ in Eq. 3.1.9 to obtain the last equation given by

$$a_2 + b_2 = 0$$

These four equations give $a_1 = 4$, $b_1 = -2$, $a_2 = 4$, $b_2 = -4$. The complete solution is thus

$$\begin{bmatrix} v \\ i \end{bmatrix} = \begin{bmatrix} -8 \\ 6 \end{bmatrix} e^{-1.5t} + \begin{bmatrix} 4 \\ -2 \end{bmatrix} e^{-t} + \begin{bmatrix} 4 \\ -4 \end{bmatrix} e^{-2t} \qquad (3.1.13)$$

The Procedure

In the above example the roots to the characteristic equation were distinct. The procedure is different if two or more roots are identical. We first consider the case of distinct roots.

For a type A nth order system the state equation is of the form

$$\dot{x}(t) = Ax(t) + Bq(t) \qquad (3.1.14)$$

where $q(t)$ is an rx1 vector called the forcing function, $x(t)$ is an nx1 vector called the state vector, A is an $n \times n$ constant matrix called the system matrix, and B is an $n \times r$ constant matrix.

The solution of Eq. 3.1.14 is given by

$$x(t) = x_1 e^{s_1 t} + \cdots + x_n e^{s_n t} + x_p(t) \qquad (3.1.15)$$

where x_1, ..., x_n are the coefficient vectors of the homogeneous solution and $x_p(t)$ is the particular solution, an $n \times 1$ vector determined by the forcing function. For example, in Eq. 3.1.11 x_1 and x_2 are given by

$$x_1 = \begin{bmatrix} a_1 \\ b_1 \end{bmatrix}, \quad x_2 = \begin{bmatrix} a_2 \\ b_2 \end{bmatrix}$$

Substituting Eq. 3.1.15 into 3.1.14 gives

$$\dot{x}(t) = Ax_1 e^{s_1 t} + \cdots + Ax_n e^{s_n t} + Ax_p(t) + Bq(t) \qquad (3.1.16)$$

Take the derivative of Eq. 3.1.15 to get

$$\dot{x}(t) = s_1 x_1 e^{s_1 t} + \cdots + s_n x_n e^{s_n t} + \dot{x}_p(t) \qquad (3.1.17)$$

Now set Eq. 3.1.16 equal to Eq. 3.1.17 to obtain

$$Ax_p(t) + Bq(t) - \dot{x}_p(t) = (s_1 I - A)x_1 e^{s_1 t} + \cdots + (s_n I - A)x_n e^{s_n t} \qquad (3.1.18)$$

3.7

The term left of the equal sign is zero because the particular solution must satisfy Eq. 3.1.14. Thus the sum on the right must equal zero. Furthermore, each term by itself is zero because each of the homogeneous solutions must satisfy Eq. 3.1.14. This plus the initial conditions is enough information to solve for the coefficients $x_1, x_2, ..., x_n$. All this leads to the following equations.

$$x(0) = x_1 + \cdots + x_n + x_p(0) \tag{3.1.19}$$

$$(s_1 I - A)x_1 = 0 \tag{3.1.20a}$$

$$\vdots \quad \vdots$$

$$(s_n I - A)x_n = 0 \tag{3.1.20n}$$

There are n^2 coefficients to find, and we obtain n equations from the initial conditions, Eq. 3.1.19. The rank of each $(s_i I - A)$ matrix is $n - 1$, and we obtain the other $n(n - 1)$ equations from the n equations, Eq. 3.1.20a through 3.1.20n.

Problem 3.1.1. Solve the state equations for the circuit in Fig. 3.1.2. The system is initially at rest.

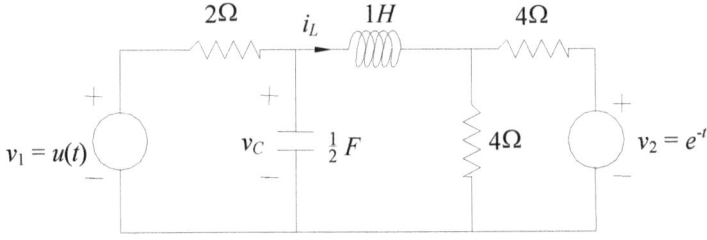

Fig. 3.1.2

Solution: The state equations are given by (see problem 2.1.3 of Chapter 2.)

$$\frac{dv_c}{dt} = -v_c - 2i_L + v_1$$

$$\frac{di_L}{dt} = v_c - 2i_L - \frac{1}{2}v_2$$

To simplify notation, let $v = v_C$ and $i = i_L$. In matrix form these equations become

$$\begin{bmatrix} \dot{v} \\ \dot{i} \end{bmatrix} = \begin{bmatrix} -1 & -2 \\ 1 & -2 \end{bmatrix}\begin{bmatrix} v \\ i \end{bmatrix} + \begin{bmatrix} 1 & 0 \\ 0 & -\frac{1}{2} \end{bmatrix}\begin{bmatrix} u(t) \\ e^{-t} \end{bmatrix} \qquad (A)$$

Particular Solution

The particular solution has the form of the forcing functions plus all derivatives. Therefore assume

$$\begin{bmatrix} v_p \\ i_p \end{bmatrix} = \begin{bmatrix} k_1 + k_3 e^{-t} \\ k_2 + k_4 e^{-t} \end{bmatrix}$$

The derivative is given by

$$\begin{bmatrix} \dot{v}_p \\ \dot{i}_p \end{bmatrix} = \begin{bmatrix} -k_3 e^{-t} \\ -k_4 e^{-t} \end{bmatrix}$$

Substitute these into the state equation (A) to obtain

$$\begin{bmatrix} -k_3 e^{-t} \\ -k_4 e^{-t} \end{bmatrix} = \begin{bmatrix} -1 & -2 \\ 1 & -2 \end{bmatrix}\begin{bmatrix} k_1 + k_3 e^{-t} \\ k_2 + k_4 e^{-t} \end{bmatrix} + \begin{bmatrix} 1 & 0 \\ 0 & -\frac{1}{2} \end{bmatrix}\begin{bmatrix} 1 \\ e^{-t} \end{bmatrix}$$

Equating like coefficients gives

$$\begin{bmatrix} 0 \\ 0 \end{bmatrix} = \begin{bmatrix} -k_1 - 2k_2 \\ k_1 - 2k_2 \end{bmatrix} + \begin{bmatrix} 1 \\ 0 \end{bmatrix}$$

and

3.9

$$\begin{bmatrix} -k_3 \\ -k_4 \end{bmatrix} = \begin{bmatrix} -k_3 - 2k_4 \\ k_3 - 2k_4 \end{bmatrix} + \begin{bmatrix} 0 \\ -\frac{1}{2} \end{bmatrix}$$

This gives the particular solution:

$$\begin{bmatrix} v_p \\ i_p \end{bmatrix} = \begin{bmatrix} \frac{1}{2} + \frac{1}{2}e^{-t} \\ \frac{1}{4} \end{bmatrix}$$

Homogeneous Solution

Set the forcing function equal to zero to obtain the homogeneous equation:

$$\begin{bmatrix} \dot{v}_h \\ \dot{i}_h \end{bmatrix} = \begin{bmatrix} -1 & -2 \\ 1 & -2 \end{bmatrix} \begin{bmatrix} v_h \\ i_h \end{bmatrix} \qquad \text{(B)}$$

The solution must be exponential, so assume

$$\begin{bmatrix} v_h \\ i_h \end{bmatrix} = \begin{bmatrix} a \\ b \end{bmatrix} e^{st}$$

The derivative is given by

$$\begin{bmatrix} \dot{v}_h \\ \dot{i}_h \end{bmatrix} = \begin{bmatrix} sa \\ sb \end{bmatrix} e^{st} = \begin{bmatrix} s & 0 \\ 0 & s \end{bmatrix} \begin{bmatrix} a \\ b \end{bmatrix} e^{st}$$

Substitute this into the homogeneous equation (B).

$$\begin{bmatrix} (s+1) & 2 \\ -1 & (s+2) \end{bmatrix} \begin{bmatrix} a \\ b \end{bmatrix} = \begin{bmatrix} 0 \\ 0 \end{bmatrix} \qquad \text{(C)}$$

The characteristic equation is therefore

$$\begin{vmatrix} (s+1) & 2 \\ -1 & (s+2) \end{vmatrix} = s^2 + 3s + 4 = 0$$

Or

$$s = -\frac{3}{2} + j\frac{\sqrt{7}}{2}, \quad -\frac{3}{2} - j\frac{\sqrt{7}}{2}$$

The homogeneous solution is thus

$$\begin{bmatrix} v_h \\ i_h \end{bmatrix} = \begin{bmatrix} a_1 \\ b_1 \end{bmatrix} e^{-\left(\frac{3}{2} - j\frac{\sqrt{7}}{2}\right)t} + \begin{bmatrix} a_2 \\ b_2 \end{bmatrix} e^{-\left(\frac{3}{2} + j\frac{\sqrt{7}}{2}\right)t}$$

The total solution is the sum of the particular and homogeneous solutions.

$$\begin{bmatrix} v \\ i \end{bmatrix} = \begin{bmatrix} \frac{1}{2} + \frac{1}{2}e^{-t} \\ \frac{1}{4} \end{bmatrix} + \begin{bmatrix} a_1 \\ b_1 \end{bmatrix} e^{-\left(\frac{3}{2} - j\frac{\sqrt{7}}{2}\right)t} + \begin{bmatrix} a_2 \\ b_2 \end{bmatrix} e^{-\left(\frac{3}{2} + j\frac{\sqrt{7}}{2}\right)t}$$

Now solve for the four constants a_1, a_2, b_1, b_2. The initial conditions give us two equations:

$$\begin{bmatrix} 0 \\ 0 \end{bmatrix} = \begin{bmatrix} 1 \\ \frac{1}{4} \end{bmatrix} + \begin{bmatrix} a_1 \\ b_1 \end{bmatrix} + \begin{bmatrix} a_2 \\ b_2 \end{bmatrix}$$

Equation (C) with $s = s_1$ gives

$$-a_1 + \left(\frac{1}{2} + j\frac{\sqrt{7}}{2}\right)b_1 = 0$$

and with $s = s_2$ we get

$$-a_2 + \left(\frac{1}{2} - j\frac{\sqrt{7}}{2}\right)b_2 = 0$$

Of course the complex numbers complicate the remaining algebra somewhat, but the solution of these four equations yields

$$a_1 = 1 - j\frac{1}{4}\left(\frac{3}{\sqrt{7}} - \sqrt{7}\right)$$

$$a_2 = -2 + j\frac{1}{4}\left(\frac{3}{\sqrt{7}} - \sqrt{7}\right)$$

$$b_1 = \frac{1}{2} - j\frac{3}{2\sqrt{7}}$$

3.11

$$b_2 = -\frac{3}{4} + j\frac{3}{2\sqrt{7}}$$

Therefore the solution is given by

$$
\begin{bmatrix} v \\ i \end{bmatrix} = \begin{bmatrix} \frac{1}{2} + \frac{1}{2}e^{-t} \\ \frac{1}{4} \end{bmatrix} + \begin{bmatrix} 1 - j\frac{1}{4}\left(\frac{3}{\sqrt{7}} - \sqrt{7}\right) \\ \frac{1}{2} - j\frac{3}{2\sqrt{7}} \end{bmatrix} e^{-\left(\frac{3}{2} - j\frac{\sqrt{7}}{2}\right)t}
$$
$$
+ \begin{bmatrix} -2 + j\frac{1}{4}\left(\frac{3}{\sqrt{7}} - \sqrt{7}\right) \\ -\frac{3}{4} + j\frac{3}{2\sqrt{7}} \end{bmatrix} e^{-\left(\frac{3}{2} + j\frac{\sqrt{7}}{2}\right)t}
$$

Problem 3.1.2. Solve the state equations for the circuit in Fig. 3.1.3. The system is initially at rest.

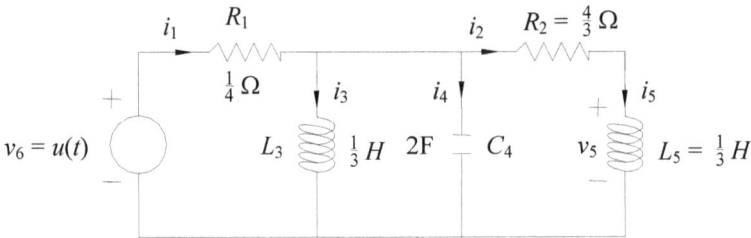

Fig. 3.1.3

Solution. The state equations are given by

$$
\begin{bmatrix} \dot{i}_3 \\ \dot{v}_4 \\ \dot{i}_5 \end{bmatrix} = \begin{bmatrix} 0 & 3 & 0 \\ -\frac{1}{2} & -2 & -\frac{1}{2} \\ 0 & 3 & -4 \end{bmatrix} \begin{bmatrix} i_3 \\ v_4 \\ i_5 \end{bmatrix} + \begin{bmatrix} 0 \\ 2 \\ 0 \end{bmatrix} u(t)
\qquad (A)
$$

with all three state variables having an initial value of zero.

Particular Solution: Since the forcing function is constant for $t > 0$ the particular solution is given by

$$\begin{bmatrix} i_3 \\ v_4 \\ i_5 \end{bmatrix} = \begin{bmatrix} k_1 \\ k_2 \\ k_3 \end{bmatrix}$$

Substitute this into the state equation (A) to get

$$\begin{bmatrix} 0 \\ 0 \\ 0 \end{bmatrix} = \begin{bmatrix} 0 & 3 & 0 \\ -\frac{1}{2} & -2 & -\frac{1}{2} \\ 0 & 3 & -4 \end{bmatrix} \begin{bmatrix} k_1 \\ k_2 \\ k_3 \end{bmatrix} + \begin{bmatrix} 0 \\ 2 \\ 0 \end{bmatrix}$$

from which we find $k_1 = 4$, $k_2 = k_3 = 0$. So the particular solution is given by

$$\begin{bmatrix} i_3 \\ v_4 \\ i_5 \end{bmatrix} = \begin{bmatrix} 4 \\ 0 \\ 0 \end{bmatrix} \qquad \text{(B)}$$

Homogeneous Solution. Assume the homogeneous solution is exponential.

$$\begin{bmatrix} i_3 \\ v_4 \\ i_5 \end{bmatrix} = \begin{bmatrix} a \\ b \\ c \end{bmatrix} e^{st}$$

The derivative vector is therefore given by

$$\begin{bmatrix} \dot{i}_3 \\ \dot{v}_4 \\ \dot{i}_5 \end{bmatrix} = \begin{bmatrix} s & 0 & 0 \\ 0 & s & 0 \\ 0 & 0 & s \end{bmatrix} \begin{bmatrix} a \\ b \\ c \end{bmatrix} e^{st}$$

Substitute this into Eq. (A) with the forcing function set equal to zero to obtain

3.13

$$\begin{bmatrix} s & 0 & 0 \\ 0 & s & 0 \\ 0 & 0 & s \end{bmatrix} \begin{bmatrix} a \\ b \\ c \end{bmatrix} = \begin{bmatrix} 0 & 3 & 0 \\ -\frac{1}{2} & -2 & -\frac{1}{2} \\ 0 & 3 & -4 \end{bmatrix} \begin{bmatrix} a \\ b \\ c \end{bmatrix}$$

or

$$\begin{bmatrix} s & -3 & 0 \\ \frac{1}{2} & s+2 & \frac{1}{2} \\ 0 & -3 & s+4 \end{bmatrix} \begin{bmatrix} a \\ b \\ c \end{bmatrix} = \begin{bmatrix} 0 \\ 0 \\ 0 \end{bmatrix} \qquad (C)$$

The characteristic equation is

$$\begin{vmatrix} s & -3 & 0 \\ \frac{1}{2} & s+2 & \frac{1}{2} \\ 0 & -3 & s+4 \end{vmatrix} = s^3 + 6s^2 + 11s + 6 = 0$$

From which $s = -1, -2, -3$.

The complete solution is given by

$$\begin{bmatrix} i_3 \\ v_4 \\ i_5 \end{bmatrix} = \begin{bmatrix} 4 \\ 0 \\ 0 \end{bmatrix} + \begin{bmatrix} a_1 \\ b_1 \\ c_1 \end{bmatrix} e^{-t} + \begin{bmatrix} a_2 \\ b_2 \\ c_2 \end{bmatrix} e^{-2t} + \begin{bmatrix} a_3 \\ b_3 \\ c_3 \end{bmatrix} e^{-3t} \qquad (D)$$

Now solve for the nine constants a_1 through c_3. Substitute the initial conditions into the complete solution, Eq. (D), to obtain three independent equations.

$$0 = 4 + a_1 + a_2 + a_3$$

$$0 = 0 + b_1 + b_2 + b_3$$

$$0 = 0 + c_1 + c_2 + c_3$$

The rank of the matrix in Eq. (C) is two. Therefore we can obtain two linearly independent equations for each value of s from this equation. This gives us the other six equations. For $s = -1$

3.14

$$a_1 = -3b_1$$

$$b_1 = c_1$$

For $s = -2$

$$a_2 + c_2 = 0$$

$$-3b_2 + 2c_2 = 0$$

Finally, for $s = -3$

$$a_3 + b_3 = 0$$

$$3b_3 = c_3$$

After solving these 9 equations for the 9 unknowns the complete solution is given by

$$\begin{bmatrix} i_3 \\ v_4 \\ i_5 \end{bmatrix} = \begin{bmatrix} 4 \\ 0 \\ 0 \end{bmatrix} + \begin{bmatrix} -9 \\ 3 \\ 3 \end{bmatrix} e^{-t} + \begin{bmatrix} 6 \\ -4 \\ -6 \end{bmatrix} e^{-2t} + \begin{bmatrix} -1 \\ 1 \\ 3 \end{bmatrix} e^{-3t}$$

3.1.3. Repeated Roots

If you recall, the discussion in Section 3.1.2 above was only for distinct roots of the characteristic equation. The solution is somewhat different if one or more roots are repeated. We will discuss only the homogeneous equation here, since the particular solution adds nothing. The homogeneous state equation is given by

$$\dot{x}(t) = Ax(t), \quad x(0) = x_0 \qquad (3.1.21)$$

Suppose, first of all, that this is a second-order system with repeated roots. Then the solution is given by

$$x(t) = x_1 e^{s_1 t} + x_2 t e^{s_1 t} \qquad (3.1.22)$$

as can be shown by direct substitution. Now substitute Eq. 3.1.22 into 3.1.21 to obtain

$$\dot{x}(t) = Ax_1 e^{s_1 t} + Ax_2 t e^{s_1 t} \tag{3.1.23}$$

And if we take the derivative of Eq. 3.1.22 we obtain

$$\dot{x}(t) = s_1 x\_1\, e^{s_1 t} + x_2 (e^{s_1 t} + t s_1 e^{s_1 t}) \tag{3.1.24}$$

Setting Eq. 3.1.23 equal to 3.1.24 gives

$$(s_1 I x_1 + I x_2 - Ax_1) e^{s_1 t} + (s_1 I x_2 - Ax_2) t e^{s_1 t} = 0 \tag{3.1.25}$$

This equation must hold for any time $t \geq 0$. In particular, at $t = 0$ the right hand term becomes zero, which gives

$$x_2 = (A - s_1 I) x_1 \tag{3.1.26}$$

Since the left-hand term is zero in Eq. 3.1.25, the right-hand term must also be zero. This gives

$$(A - I s_1) x_2 = 0 \tag{3.1.27}$$

Now substitute the value of x_2 from Eq. 3.1.26 into 3.1.27 to obtain

$$(A - I s_1)^2 x_1 = 0 \tag{3.1.28}$$

but

$$(A - I s_1)^2 = 0 \tag{3.1.29}$$

for a second-order system. (This can be shown in either of two ways, by applying Jordan's lemma, or by the Cayley-Hamilton theorem. For example, the Cayley-Hamilton theorem states that any *nxn* matrix A satisfies its own characteristic equation, and Eq. 3.1.29 is this equation.)

Since Eq. 3.1.29 is true, x_1 in Eq. 3.1.28 is arbitrary. Therefore choose x_1 to satisfy the initial conditions in Eq. 3.1.22, then solve for x_2 from Eq. 3.1.26.

For a third or higher-order system the situation becomes somewhat more complicated because Eq. 3.1.29 is no longer valid. Since repeated roots are rare to begin with, and since it would take too long to complete our discussion, we will stop here.

Problem 3.1.3. Solve the state equations for the circuit in Fig. 3.1.4. All initial conditions are zero.

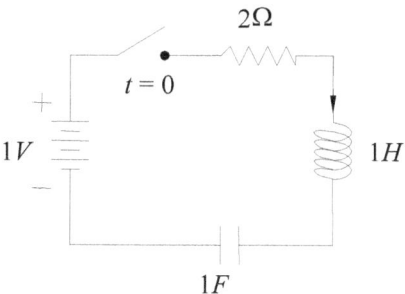

Fig. 3.1.4

Solution: The state equations are given by

$$\begin{bmatrix} \dot{i} \\ \dot{v} \end{bmatrix} = \begin{bmatrix} -2 & -1 \\ 1 & 0 \end{bmatrix}\begin{bmatrix} i \\ v \end{bmatrix} + \begin{bmatrix} 1 \\ 0 \end{bmatrix}, \quad \begin{bmatrix} i(0) \\ v(0) \end{bmatrix} = \begin{bmatrix} 0 \\ 0 \end{bmatrix} \qquad \text{(A)}$$

where $i(t)$ is the inductor current and $v(t)$ is the capacitor voltage.

<u>Particular Solution:</u> The particular solution is given by

$$\begin{bmatrix} i_p \\ v_p \end{bmatrix} = \begin{bmatrix} k_1 \\ k_2 \end{bmatrix} = \begin{bmatrix} 0 \\ 1 \end{bmatrix}$$

which is obtained by substitution into Eq. (A).

<u>Homogeneous Solution:</u> The characteristic equation is

$$\left|(sI - A)\right| = \begin{vmatrix} s+2 & 1 \\ -1 & s \end{vmatrix} = s^2 + 2s + 1 = 0$$

which has repeated roots, $s = -1, -1$. Therefore the homogeneous solution is

3.17

$$\begin{bmatrix} i_h \\ v_h \end{bmatrix} = \begin{bmatrix} a_1 \\ b_1 \end{bmatrix} e^{-t} + \begin{bmatrix} a_2 \\ b_2 \end{bmatrix} t e^{-t}$$

The total solution is the sum of the particular and homogeneous solution, given by

$$\begin{bmatrix} i \\ v \end{bmatrix} = \begin{bmatrix} 0 \\ 1 \end{bmatrix} + \begin{bmatrix} a_1 \\ b_1 \end{bmatrix} e^{-t} + \begin{bmatrix} a_2 \\ b_2 \end{bmatrix} t e^{-t} \tag{B}$$

Find x_1 by substituting the initial conditions into Eq. (B).

$$\begin{bmatrix} 0 \\ 0 \end{bmatrix} = \begin{bmatrix} 0 \\ 1 \end{bmatrix} + \begin{bmatrix} a_1 \\ b_1 \end{bmatrix}$$

or $a_1 = 0$, $b_1 = -1$. Now use Eq. 3.1.26 with $s_1 = -1$ to obtain

$$x_2 = \begin{bmatrix} a_2 \\ b_2 \end{bmatrix} = \begin{bmatrix} -1 & -1 \\ 1 & 1 \end{bmatrix} \begin{bmatrix} 0 \\ 1 \end{bmatrix}$$

or $a_2 = 1$, $b_2 = -1$. The complete solution is

$$\begin{bmatrix} i \\ v \end{bmatrix} = \begin{bmatrix} 0 \\ 1 \end{bmatrix} + \begin{bmatrix} 0 \\ -1 \end{bmatrix} e^{-t} + \begin{bmatrix} 1 \\ -1 \end{bmatrix} t e^{-t}$$

Self Test, Objective 3.1.

Find the closed-form analytical solution of the state equations for the network in Fig.3.1.5. All initial conditions are zero and the state equations are given by

$$\begin{bmatrix} \dot{v}_2 \\ \dot{i}_3 \end{bmatrix} = \begin{bmatrix} -4 & -2 \\ 1 & -1 \end{bmatrix} \begin{bmatrix} v_2 \\ i_3 \end{bmatrix} + \begin{bmatrix} 4 \\ 0 \end{bmatrix}, \quad t \geq 0$$

v_2 is the capacitor voltage and i_3 is the inductor current.

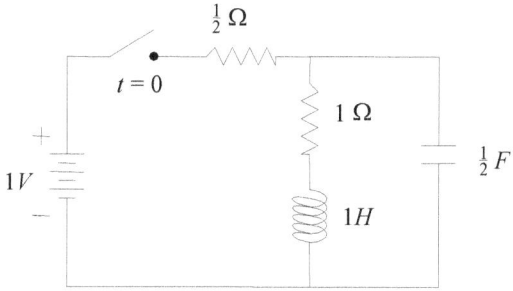

Fig. 3.1.5

Pre-Test Answers:

1. $x_1 = \frac{1}{4}$, $x_2 = -7/4$
2. $|2I - A| = 4$

Self Test Answers:

Objective 3.1.

$$\begin{bmatrix} v_2 \\ i_3 \end{bmatrix} = \begin{bmatrix} 2 \\ -2 \end{bmatrix} e^{-2t} + \begin{bmatrix} -\dfrac{8}{3} \\ \dfrac{4}{3} \end{bmatrix} e^{-3t} + \begin{bmatrix} \dfrac{2}{3} \\ \dfrac{2}{3} \end{bmatrix}, \quad t \geq 0$$

Chapter 4

The Solution of State Equations - Type B Networks

Objectives: After completing this chapter you should be able to do the following:

4.1. Find a closed-form analytical solution for type B networks by impulse matching.

4.2. Find a closed-form analytical solution for type B networks by reducing the form to type A equations.

Rationale.

This chapter presents two methods of solving type B equations. The first method (objective 4.1) provides more insight into the nature of initial conditions, while the second method (objective 4.2) will allow us to use one form in future work for type A or B networks.

Objective 4.1. Find a closed-form analytical solution for type B networks by impulse matching.

In Chapter 3 we paid scant attention to initial conditions. There was no need to because all state variables were continuous. That is, if there are no impulses in the forcing function, then the state variables are continuous at every instant of time. Thus when we specify an initial condition at $t = 0$ it matters not whether we mean at $t=0^-$ or at $t = 0^+$. This is a result of the continuity theorem.

The Continuity Theorem: The solution to the state equation 4.1.1 (below) with initial condition $x(0) = x_0$ is continuous for all time if the forcing function $f(t)$ contains at most a finite number of finite discontinuities.

$$\dot{x}(t) = Ax(t) + f(t) \tag{4.1.1}$$

Here the forcing function is $f(t)$. For type A networks $f(t) = Bq(t)$. For type B networks $f(t) = Bq(t) + E\dot{q}(t)$. Therefore type A networks satisfy the continuity theorem since all current and voltage sources cannot have impulses and higher-order singularity functions. For type B networks, however, if a source contains a step discontinuity (say a switch closing) then the derivative of $q(t)$ will contain an impulse, and the continuity theorem no longer applies to this case.

The validity of the continuity theorem can be seen from the following argument. Suppose that one of the state variables $x_k(t)$ contains a step discontinuity at $t = t_1$. Then the derivative $\dot{x}_k(t)$ will be infinite at $t = t_1$. But the right-hand side of Eq. 4.1.1 is finite at $t = t_1$ if $f(t)$ satisfies the continuity theorem. Hence, we arrive at a contradiction.

You can see that for type B networks, if the forcing function contains discontinuities and that particular forcing function appears in $\dot{q}(t)$, then at least some state variables must contain discontinuities. That is, if the right-hand side of Eq. 4.1.1 contains an impulse, then the left side must also. Since the left side is the derivative of state variables, then the state variables must be discontinuous. Our purpose in this objective is to arrive at a method of handling these discontinuities.

Impulse Matching

For type B networks the solution is identical to that for type A networks up to the point of using initial conditions. At this point the continuity theorem no longer applies, so we cannot use the same methods to solve for the coefficients in the homogeneous solution. Here is an example to illustrate the method.

The circuit in Fig. 4.1.1 is from Chapter 2 where the state equation is given by

$$\dot{v}(t) = -\tfrac{1}{3}v(t) + \tfrac{1}{6}v_s(t) + \tfrac{1}{3}\dot{v}_s(t)$$

where $v(t)$ is the voltage across C_1. Let us assume that $v(0^-) = 0$ and that $v_s(t)$ is a unit step voltage. Then $\dot{v}_s(t) = \delta(t)$. The equation we must solve is therefore given by

$$\dot{v}(t) = -\tfrac{1}{3}v(t) + \tfrac{1}{6}v_s(t) + \tfrac{1}{3}\delta(t), \quad v(0^-) = 0 \qquad (4.1.2)$$

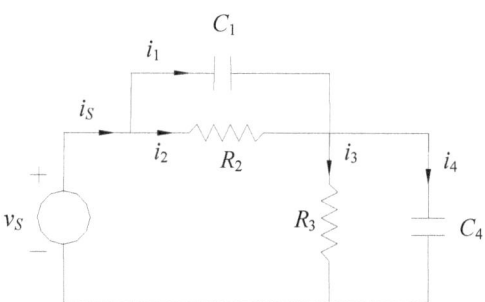

Fig. 4.1.1.

Since the right side of Eq. 4.1.2 contains a delta function, the left side must also. Thus $\dot{v}(t)$ contains an impulse of strength $1/3$ at $t = 0$. That is, $v(0^+) = v(0^-) + 1/3$. With this in mind let us now solve Eq. 4.1.2. For $t > 0$ the δ-function is eliminated and Eq. 4.1.2 becomes

$$\dot{v}(t) = -\frac{1}{3}v(t) + \frac{1}{6}v_s(t) \tag{4.1.3}$$

The particular solution is proportional to the forcing function plus all its derivatives. In Eq. 4.1.3 the forcing function is $1/6$, so the particular solution is a constant k. Substitute this into Eq. 4.1.3 to obtain the particular solution given by

$$v_p(t) = \frac{1}{2}, \qquad t > 0$$

The homogeneous solution is

$$v_h(t) = ae^{-\frac{1}{3}t}, \qquad t > 0$$

so the total solution is given by

$$v(t) = \frac{1}{2} - ae^{-\frac{1}{3}t}, \ t > 0 \tag{4.1.4}$$

We have already discovered that $v(0^+) = 1/3$, so substituting this into Eq. 4.1.4 gives

$$v(t) = \frac{1}{2} - \frac{1}{6}e^{-\frac{1}{3}t}, \ t > 0 \tag{4.1.5}$$

This example gives the general procedure for solving type B state equations by impulse matching. For the general equation given by

$$\dot{x}(t) = Ax(t) + Bq(t) + E\dot{q}(t), \quad t \geq 0 \qquad (4.1.6)$$

where the forcing functions $q(t)$ contain at most step discontinuities, and where the initial values of $x(t)$ at $t = 0^-$ are known, the procedure is as follows:

For any of the n equations in Eq. 4.1.6 if there are no derivative terms of the forcing function, then the continuity theorem implies that $x_i(0^-) = x_i(0^+)$. If, however, an impulse appears in the derivative of the forcing function of strength H, then the derivative of the state variable $\dot{x}_i(t)$ must also contain an impulse of strength H. Therefore $x_i(t)$ must contain a step discontinuity at $t = 0$ of height H. That is,

$$x_i(0^+) = x_i(0^-) + H \qquad (4.1.7)$$

Now rewrite Eq. 4.1.6 for $t > 0$ to eliminate the delta functions and solve this equation in the usual manner, because now it is a type A equation.

Problem 4.1.1. Find the current in resistor R_1 in Fig. 4.1.2. The source and initial inductor currents are

$$i_s(t) = e^{-t}u(t), \quad i_2(0^-) = 1, \quad i_3(0^-) = -1$$

Solution: Problem 2.1.4 solution in Chapter 2 used i_2 as the state equation, given by

$$(L_1 + L_3)\frac{di_2(t)}{dt} = -R_1 i_2(t) + L_3 \frac{di_s}{dt}$$

With the values $L_1 = L_2 = 1H$ and $R = 1\Omega$, this state equation becomes

$$\frac{di_2(t)}{dt} = -\frac{1}{2}i_2(t) + \frac{1}{2}\frac{di_s(t)}{dt} = -\frac{1}{2}i_2(t) + \frac{1}{2}\delta(t) - \frac{1}{2}e^{-t}, \quad t \geq 0 \qquad (A)$$

First determine $i_2(0^+)$ from $i_2(0^-)$. Since there is a delta function of strength ½ at $t = 0$ on the right side of Eq. (A), the left side must contain an

4.4

impulse of strength ½. Therefore $i_2(t)$ has a step of height ½ at $t = 0$. Apply this in Eq. 4.1.7 to obtain

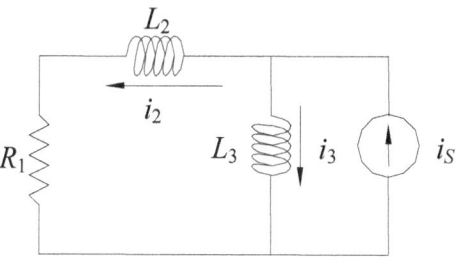

Fig. 4.1.2.

$$i_2(0^+) = i_2(0^-) + \tfrac{1}{2} = 1.5 \tag{B}$$

Next rewrite Eq. (A) for $t > 0$.

$$\frac{di_2(t)}{dt} = -\frac{1}{2}i_2(t) - \frac{1}{2}e^{-t}, \qquad t > 0$$

The particular solution is given by

$$i_p(t) = e^{-t}, \qquad t > 0$$

and the homogeneous solution is

$$i_h(t) = ae^{-\frac{1}{2}t}$$

Therefore the total solution is

$$i_2(t) = e^{-t} + ae^{-\frac{1}{2}t}$$

Substituting initial conditions from (B) gives

$$i_2(t) = e^{-t} + \frac{1}{2}e^{-\frac{1}{2}t}, \qquad t > 0$$

Problem 4.1.2. The circuit in Fig. 4.1.3 is taken from problem 2.1.5 of Chapter 2. Find $v_3(t)$ and $i_4(t)$ if the initial conditions are

$$v_3(0^-) = 0, \quad i_4(0^-) = 0.$$

The sources are $v_1(t) = e^{-t}u(t), \quad i_6(t) = u(t)$

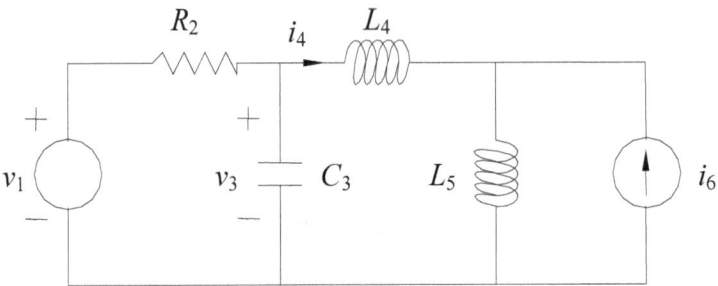

Fig. 4.1.3.

Solution: The state equations are given by

$$\dot{v}_3(t) = -5v_3(t) - i_4(t) + 5e^{-t}, \quad t \geq 0$$

$$i_4(t) = 6v_3(t) - \tfrac{1}{2}\delta(t), \quad t \geq 0$$

or, in matrix form

$$\begin{bmatrix} \dot{v}_3 \\ \dot{i}_4 \end{bmatrix} = \begin{bmatrix} -5 & -1 \\ 6 & 0 \end{bmatrix} \begin{bmatrix} v_3 \\ i_4 \end{bmatrix} + \begin{bmatrix} 5 & 0 \\ 0 & 0 \end{bmatrix} \begin{bmatrix} e^{-t} \\ 1 \end{bmatrix}$$

$$+ \begin{bmatrix} 0 & 0 \\ 0 & -\tfrac{1}{2} \end{bmatrix} \begin{bmatrix} \delta(t) - e^{-t} \\ \delta(t) \end{bmatrix}$$

Now to find the initial conditions at $t = 0^+$ use Eq. 4.1.7 to obtain

$$v_3(0^+) = v_3(0^-) = 0$$

$$i_4(0^+) = i_4(0^-) - \tfrac{1}{2} = -\tfrac{1}{2}$$

Next, rewrite the state equations for t > 0.

$$\begin{bmatrix} \dot{v}_3 \\ \dot{i}_4 \end{bmatrix} = \begin{bmatrix} -5 & -1 \\ 6 & 0 \end{bmatrix} \begin{bmatrix} v_3 \\ i_4 \end{bmatrix} + \begin{bmatrix} 5e^{-t} \\ 0 \end{bmatrix}, \quad t > 0 \tag{A}$$

The particular solution (absent the details) of Eq. (A) is

$$\begin{bmatrix} v_3 \\ i_4 \end{bmatrix} = \begin{bmatrix} k_1 \\ k_2 \end{bmatrix} e^{-t} = \begin{bmatrix} -\tfrac{5}{2} \\ 15 \end{bmatrix} e^{-t}$$

For the homogeneous solution, first find the roots of the characteristic equation.

$$|sI - A| = \begin{vmatrix} s+5 & 1 \\ -6 & s \end{vmatrix} = s^2 + 5s + 6 = 0$$

or $s = -2, -3$. Therefore the homogeneous solution is

$$\begin{bmatrix} v_3 \\ i_4 \end{bmatrix} = \begin{bmatrix} a_1 \\ b_1 \end{bmatrix} e^{-2t} + \begin{bmatrix} a_2 \\ b_2 \end{bmatrix} e^{-3t}$$

Add the particular and homogeneous solutions to get the complete solution, given by

$$\begin{bmatrix} v_3 \\ i_4 \end{bmatrix} = \begin{bmatrix} a_1 \\ b_1 \end{bmatrix} e^{-2t} + \begin{bmatrix} a_2 \\ b_2 \end{bmatrix} e^{-3t} + \begin{bmatrix} -\tfrac{5}{2} \\ 15 \end{bmatrix} e^{-t}$$

Use Eqs. 3.1.19 and 3.1.20 and substitute the initial conditions to solve for the constants. The final solution is given by

$$\begin{bmatrix} v_3 \\ i_4 \end{bmatrix} = \begin{bmatrix} 10.5 \\ -31.5 \end{bmatrix} e^{-2t} + \begin{bmatrix} -8 \\ 16 \end{bmatrix} e^{-3t} + \begin{bmatrix} -2.5 \\ 15 \end{bmatrix} e^{-t}$$

Self Test, Objective 4.1.

Find the closed form analytical solution of the state equation for the network in Fig. 4.1.4. Use the impulse matching method. The sources are

$$v_1(t) = e^{-t}u(t), \quad i_2(t) = u(t)$$

All initial conditions are zero and the state equation is given by

$$\dot{v}_5 = -\tfrac{4}{3}v_5 + \tfrac{2}{3}v_1 + \tfrac{2}{3}i_2 + \tfrac{2}{3}\dot{v}_1$$

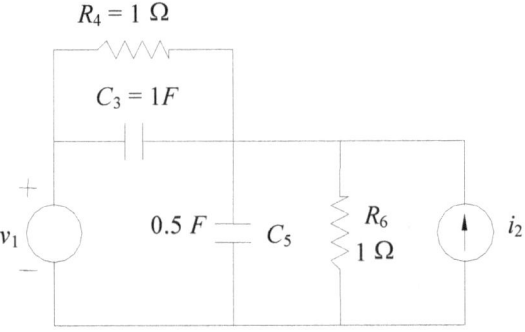

Fig. 4.1.4.

Objective 4.2. Find the closed form analytical solution for type B networks by reducing the form to type A equations.

The linear state equations for type B networks are of the form given by

$$\dot{x} = Ax + Bq + E\dot{q} \tag{4.2.1}$$

This system of equations can always be reduced to the form for type A equations by a change of variable. Let

$$z = x - Eq \tag{4.2.2}$$

4.8

Substitute Eq. 4.2.2 into 4.2.1 to get

$$\dot{z} = Az + (B + AE)q \qquad (4.2.3)$$

This is the desired form. Here is the procedure:

1. Use Eq. 4.2.2 to recast the type B equation as type A.

2. Solve Eq. 4.2.3 for the variable z. Note that the continuity theorem applies to Eq. 4.2.3. That is, $\qquad z(0^-) = x(0^-) = Eq(0^-) \qquad (4.2.4)$

3. Finally, use Eq. 4.2.2 to solve for x after finding z. That is,

$$x = z + Eq \qquad (4.2.5)$$

Problem 4.2.1. Solve the problem associated with Eq. 4.1.2 by reducing the state equation to type A. The type B equation is given by

$$\dot{v}(t) = -\tfrac{1}{3}v(t) + \tfrac{1}{6}u(t) + \tfrac{1}{3}\delta(t), \quad v(0^-) = 0 \qquad (A)$$

Solution: Identify the following terms:

$$q(t) = u(t) \quad A = -\tfrac{1}{3}$$

$$E = \tfrac{1}{3} \qquad B = \tfrac{1}{6}$$

Equation 4.2.2 gives

$$z(t) = v(t) - \tfrac{1}{3}u(t)$$

Or

$$v(t) = z(t) + \tfrac{1}{3}u(t)$$

$$\dot{v}(t) = \dot{z}(t) + \tfrac{1}{3}\delta(t)$$

Substitute these two equations into (A) to obtain

$$\dot{z}(t) = -\frac{1}{3}z(t) + \frac{1}{18}u(t), \quad z(0^-) = 0 \qquad (B)$$

Now solve Eq. (B) in the usual manner for type A equations to obtain

$$zt) = \frac{1}{6} - \frac{1}{6} e^{-\frac{1}{3}t}$$

Use Eq. 4.2.5 to solve for $x(t)$.

$$x(t) = \frac{1}{2} - \frac{1}{6} e^{-\frac{1}{3}t}, \quad t > 0 \qquad (4.2.6)$$

which agrees with Eq. 4.1.5.

Problem 4.2.2. Solve problem 4.1.2 by reducing the state equation to type A.

Solution: The equation that we must solve is given by

$$\begin{bmatrix} \dot{v}_3 \\ \dot{i}_4 \end{bmatrix} = \begin{bmatrix} -5 & -1 \\ 6 & 0 \end{bmatrix} \begin{bmatrix} v_3 \\ i_4 \end{bmatrix} + \begin{bmatrix} 5 & 0 \\ 0 & 0 \end{bmatrix} \begin{bmatrix} e^{-t} \\ 1 \end{bmatrix}$$
$$+ \begin{bmatrix} 0 & 0 \\ 0 & -\frac{1}{2} \end{bmatrix} \begin{bmatrix} \delta(t) - e^{-t} \\ \delta(t) \end{bmatrix}$$

with initial conditions given by

$$\begin{bmatrix} v_3(0^-) \\ i_4(0^-) \end{bmatrix} = \begin{bmatrix} 0 \\ 0 \end{bmatrix}$$

Identify the following matrices.

$$A = \begin{bmatrix} & \\ & \end{bmatrix}, \quad B = \begin{bmatrix} & \\ & \end{bmatrix}, \quad E = \begin{bmatrix} & \\ & \end{bmatrix},$$

$$q = \begin{bmatrix} & \\ & \end{bmatrix}, \quad (B + AE) = \begin{bmatrix} & \\ & \end{bmatrix}$$

Therefore Eq. 4.2.3 gives

$$\begin{bmatrix} \dot{z}_1 \\ \dot{z}_2 \end{bmatrix} = \begin{bmatrix} & \\ & \end{bmatrix} \begin{bmatrix} z_1 \\ z_2 \end{bmatrix} + \begin{bmatrix} & \\ & \end{bmatrix}, \quad t > 0$$

4.10

and the initial conditions are

$$\begin{bmatrix} z_1(0^-) \\ z_2(0^-) \end{bmatrix} = \begin{bmatrix} \ \\ \ \end{bmatrix}$$

Now solve for z and then x.

$$\begin{bmatrix} \dot{z}_1 \\ \dot{z}_2 \end{bmatrix} = \begin{bmatrix} -5 & -1 \\ 6 & 0 \end{bmatrix} \begin{bmatrix} z_1 \\ z_2 \end{bmatrix} + \begin{bmatrix} 5 & \frac{1}{2} \\ 0 & 0 \end{bmatrix} \begin{bmatrix} e^{-t} \\ 1 \end{bmatrix}$$

$$\begin{bmatrix} z_1(0) \\ z_2(0) \end{bmatrix} = \begin{bmatrix} 0 \\ 0 \end{bmatrix}$$

See solution to problem 4.1.2 for the final answer.

Self Test, Objective 4.2.

Find the closed form analytical solution of the state equations for the following network. Use the method of reducing the equations to type A. (The state equations are given in problem 2.1.6 of Chap. 2.) All initial conditions are zero. The sources are

$$v_1(t) = u(t), \quad i_6(t) = e^{-t}u(t)$$

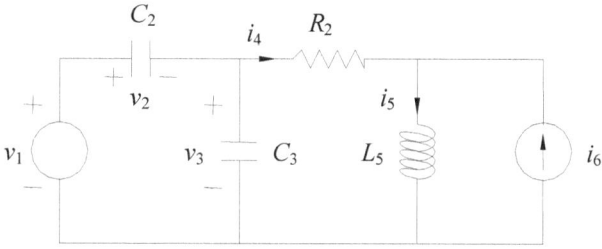

Fig. 4.2.1

Self Test Answers:

Objective 4.1:

$$v_5(t) = \frac{1}{2} + \frac{1}{6}e^{-\frac{4}{3}t}$$

Objective 4.2:

$$\begin{bmatrix} v_2 \\ i_5 \end{bmatrix} = \begin{bmatrix} -\frac{1}{2} - 14e^{-t} \\ 8e^{-t} \end{bmatrix} + \begin{bmatrix} 3.5 \\ -8 \end{bmatrix} e^{-2t} + \begin{bmatrix} 15 \\ -15 \end{bmatrix} te^{-2t}$$

Chapter 5.

Numerical Solution

Objective: Afer completing this chapter you should be able to do the following:

5.1. Write a computer program to solve approximately the state equations for both type A and type B networks.

5.1.1. First Order Equations – A Review

Here is another method of solving linear first-order differential equations. We're not talking about numerical solutions yet, this is just another way to find an analytical solution. The general form of a first-order LTI equation is given by

$$\dot{x}(t) = ax(t) + bq(t)$$

or by
$$\dot{x}(t) - ax(t) = bq(t) \tag{5.1.1}$$

Here is a useful formula: The derivative of $x(t)e^{-at}$ is given by

$$\frac{d}{dt}[x(t)e^{-at}] = [\dot{x}(t) - ax(t)]e^{-at} \tag{5.1.2}$$

Notice that the left side of Eq. 5.1.1 matches up with one of the terms in Eq. 5.1.2. This gives

$$\frac{d}{dt}[x(t)e^{-at}] = bq(t)e^{-at} \tag{5.1.3}$$

Now integrate both sides over time to obtain

$$x(t)e^{-at} = \int_{-\infty}^{t} bq(\lambda)e^{-a\lambda}d\lambda$$

$$= x(0) + \int_{0}^{t} bq(\lambda)e^{-a\lambda}d\lambda$$

So the solution for $x(t)$ is given by

$$x(t) = e^{at}x(0) + e^{at}\int_0^t bq(t)e^{-a\lambda}d\lambda \qquad (5.1.4)$$

Note: It is customary to call the first term in this solution the homogeneous solution, and the second term is called the particular integral. These terms, however, do not agree with the homogeneous and particular solutions that are found by the method of undetermined coefficients.

Problem 5.1.1. Solve the equation

$$\frac{dv}{dt} + 2v = 2e^{-t}, \quad v(0) = 0$$

Solution: The first term in Eq. 5.1.4 is zero since $v(0) = 0$. This leaves the particular integral, given by

$$v(t) = e^{-2t}\int_0^t 2e^{-\lambda}e^{2\lambda}d\lambda = 2e^{-t} - 2e^{-2t}, \quad t > 0 \qquad (5.1.5)$$

which agrees with Eq. 3.1.5.

Notice the difference in nomenclature between here and the solution in Eq. 3.1.5. Here the right side of Eq. 5.1.5 is called the particular integral, and the homogeneous solution is zero because $v(0) = 0$. In Eq. 3.1.5 the first term is called the particular solution and the second term is called the homogeneous solution. Unfortunately this inconsistent nomenclature is well established.

Series Solution:

Now we're talking numerical approximation, and the question is how to approximate the two terms in Eq. 5.1.4. The first term is an exponential. The Taylor series that defines the exponential is given by

$$e^{at} = 1 + at + \frac{a^2t^2}{2!} + \frac{a^3t^3}{3!} + \cdots \qquad (5.1.6)$$

5.2

If t is sufficiently small, then e^{at} can be approximated by the first few terms of Eq. 5.1.6. This approximation is the basis of all numerical solutions of differential equations.

Now what about the integral? First break the integration interval into small parts of length T. The simplest and least accurate approximation is to assume that $q(t)$ and e^{at} are constant over each interval, and equal to their value at the beginning of the interval. If more accuracy is desired, better methods of approximating integrals, such as Simpson's rule, can be used. But in order to simplify the present discussion we will use the above approximation.

Now consider a series of small discrete time intervals of length T. From Eq. 5.1.4 the value of x at time $t = T$ is given by

$$x(T) = e^{aT} x(0) + e^{aT} \int_0^T bq(\lambda) e^{-a\lambda} d\lambda$$

Approximate the integral by assuming that $q(\lambda) = q(0)$ and that $e^{-a\lambda} = e^0 = 1$ over the small interval $0 \leq t < T$. Therefore $x(T)$ is approximated by

$$x(T) \approx e^{aT} x(0) + T e^{aT} bq(0)$$

$$= \phi x(0) + \Delta q(0)$$

where $\phi = e^{aT}$ and $\Delta = T e^{aT} b$ are constant terms. Now let $t = 2T$. This gives

$$x(2T) = e^{2aT} x(0) + e^{2aT} \left[\int_0^T w(\lambda) d\lambda + \int_T^{2T} w(\lambda) d\lambda \right]$$

where $w(\lambda) = bq(\lambda) e^{-a\lambda}$. Now re-arrange terms

$$x(2T) = e^{aT} \left[e^{aT} x(0) + e^{aT} \int_0^T w(\lambda) d\lambda \right]$$

$$+ e^{aT} \left[e^{aT} \int_T^{2T} w(\lambda) d\lambda \right]$$

This reduces to

5.3

$$x(2T) = e^{aT}x(T) + e^{aT}\left[e^{aT} \int\limits_{T}^{2T} bq(\lambda)e^{-a\lambda}d\lambda \right]$$

Assume $e^{-a\lambda} = e^{-aT}$, $T \leq \lambda < 2T$ and that $q(\lambda)$ is constant with value $q(T)$ over this interval. This last equation then reduces to

$$x(2T) = e^{aT}x(T) + Te^{aT}bq(T)$$

$$= \phi x(T) + \Delta q(T)$$

This recursive relationship is valid for any term $t = nT$, so that the solution to the first order equation can be approximated by

$$x(nT) = \phi x[(n-1)T] + \Delta q[(n-1)T] \tag{5.1.7}$$

Note: An often used approximation is to use the definition of derivative.

$$\frac{dx}{dt} = \lim_{T \to 0} \frac{x(t+T) - x(t)}{T} \tag{5.1.8}$$

Use the right side of Eq. 5.1.8 with small T as an approximation for the derivative in

$$\dot{x}(t) = ax(t) + bq(t)$$

to arrive at a simple approximation given by

$$x(t+T) = (aT+1)x(t) + bTq(t)$$

Or $\qquad\qquad x(nT) = (aT+1)x[(n-1)T] + bTq[(n-1)T]$

Now compare this result to Eq. 5.1.7 to see that we have used only the first two terms in Eq. 5.1.6. In fact, the derivative is just a special case of the Taylor series, and that is the special case used here. This discussion reinforces our earlier remark that using the first few terms in Eq. 5.1.6 is the basis of all numerical solutions of differential equations.

Problem 5.1.2. Write a computer program to find the approximate solution to the equation

$$\frac{dv}{dt} = -2v + 2e^{-t}, \quad v(0) = 0$$

Solution: Choose increment $T = 0.1$. The term $\phi = e^{aT}$ is therefore given by

$$\phi = e^{aT} = e^{-0.2} \cong 1 - 0.2 + \frac{0.2^2}{2} - \frac{0.2^3}{6} = 0.8187$$

The Δ term is

$$\Delta = Te^{aT}b = 0.1(0.8187)2 = 0.16374$$

Thus our recursive relationship, Eq. 5.1.7, is given by

$$x(nT) = 0.8187x\big[(n-1)T\big] + 0.16374e^{-(n-1)T}$$

Figure 5.1.1 shows both the analytical solution (continuous line) and the approximate solution over $0 < t < 3$. Here is the MATLAB program that produced Fig. 5.1.1.

```
clear

t3=linspace(0,3,301);
w=2*exp(-t3)-2*exp(-2*t3);
subplot(2,1,1)
plot(t3,w,'k')
grid on
hold on
t=linspace(0,3,31);
xn=0;
tn=0;
for i=1:31
    q=exp(-tn);
    x(i)=xn;
    xn=0.8187*xn+0.16374*q;
    tn=tn+0.1;
end
stem(t,x,'k.')
grid on
```

5.5

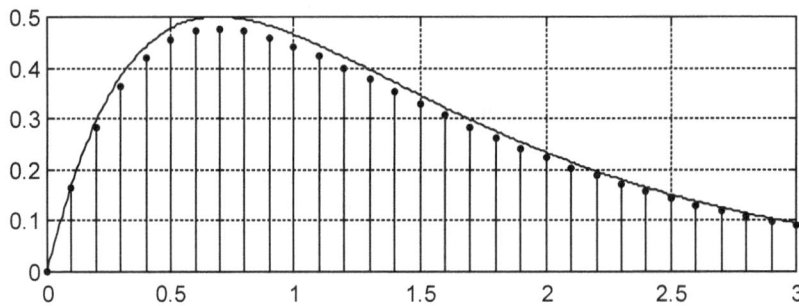

Fig. 5.1.1.

Note: The accuracy of the numerical solution given by Eq. 5.1.7 depends on (1) the magnitude of T, (2) the number of terms used in approximating ϕ, (3) the value of a, and (4) the nature of the forcing function $q(t)$.

Problem 5.1.4. Write the difference equation to find the approximate solution to the equation

$$\frac{dv}{dt} = -3v + sin(2\pi t), \quad v(0) = 0$$

Solution: Choose $T = 0.1$ and write the term $\phi = e^{aT}$.

$$\phi = e^{aT} = \underline{\hspace{2cm}}$$

Now compute Δ.

$$\Delta = Te^{aT}b = \underline{\hspace{2cm}}$$

Therefore Eq. 5.1.7 becomes

$$v(nT) = \underline{\hspace{2cm}}$$

Answers:

$$\phi = e^{-0.3} \cong 1 - 0.3 + \frac{0.3^2}{2} - \frac{0.3^3}{6} = 0.7405$$

$$\Delta = 0.1(0.7405)(1) = 0.07405$$

$$v(nT) = 0.7405 v[(n-1)T] + 0.07405 \sin[2\pi(n-1)T]$$

Numerical Solution of Type A Equations

The state equations for type A networks has the form

$$\dot{x} = Ax + Bq \tag{5.1.9}$$

The analytical solution is given by

$$x(t) = e^{At}x(0) + e^{At} \int_0^t e^{-A\lambda} BQ(\lambda) d\lambda \tag{5.1.10}$$

where e^{At} is an nxn matrix defined by the infinite series

$$e^{At} = I + At + A^2 \frac{t^2}{2!} + A^3 \frac{t^3}{3!} + \cdots \tag{5.1.11}$$

By the integral of an n component vector v we mean that each component is integrated as in Eq. 5.1.12 below.

$$\int_0^t v(\lambda) d\lambda = \begin{bmatrix} \int_0^t v_1(\lambda) d\lambda \\ \int_0^t v_2(\lambda) d\lambda \\ \vdots \\ \int_0^t v_n(\lambda) d\lambda \end{bmatrix} \tag{5.1.12}$$

Notes: (1) The solution given by Eq. 5.1.10 can be derived formally, and this is done in most textbooks that treat the subject of state equations. We will not do so here, but this solution should seem reasonable in light of Chapter 3 and Section 1 of this chapter.

(2) The first term in Eq. 5.1.10 is called the homogeneous solution, and the second term is called the particular integral. Again, these solutions are not identical to those found by the method of undetermined coefficients.

Approximate numerical solutions are found by the same methods as for first order equations. The exponential e^{At} is approximated by summing the first few terms of the series in Eq. 5.1.11, and the value of the integral is approximated by assuming that the terms are constant and equal to their value at the beginning of the interval. Thus

$$x(nT) = \phi x[(n-1)T] + \Delta Q[(n-1)T] \qquad (5.1.13)$$

where $\phi = e^{AT}$, $\Delta = Te^{AT}B$ are constant matrices.

Problem 5.1.5. Write a MATLAB program to solve the following matrix equation. Use $T = 0.1$ and approximate the exponential by the sum of the first four terms in Eq. 5.1.11.

$$\begin{bmatrix} \dot{x}_1 \\ \dot{x}_2 \end{bmatrix} = \begin{bmatrix} 0 & 2 \\ -1 & -3 \end{bmatrix} \begin{bmatrix} x_1 \\ x_2 \end{bmatrix} + \begin{bmatrix} 0 \\ 1 \end{bmatrix} e^{-1.5t}, \quad \begin{bmatrix} x_1(0) \\ x_2(0) \end{bmatrix} = \begin{bmatrix} 0 \\ 0 \end{bmatrix}$$

Solution: The first four terms in Eq. 5.1.11 are

$$\phi = \begin{bmatrix} 1 & 0 \\ 1 & 0 \end{bmatrix} + \begin{bmatrix} 0 & 2 \\ -1 & -3 \end{bmatrix}(0.1) + \begin{bmatrix} -2 & -6 \\ 3 & 7 \end{bmatrix}\frac{0.01}{2} + \begin{bmatrix} 6 & 14 \\ -7 & -15 \end{bmatrix}\frac{0.001}{6}$$

$$= \begin{bmatrix} 0.991 & 0.172 \\ -0.086 & 0.733 \end{bmatrix}$$

The matrix Δ is computed by

$$\Delta = T\phi B = 0.1 \begin{bmatrix} 0.991 & 0.172 \\ -0.086 & 0.733 \end{bmatrix} \begin{bmatrix} 0 \\ 1 \end{bmatrix} = \begin{bmatrix} 0.0172 \\ 0.0733 \end{bmatrix}$$

From Eq. 5.1.13 the numerical solution is given by

$$\begin{bmatrix} x_1(n+1) \\ x_2(n+1) \end{bmatrix} = \begin{bmatrix} 0.991 & 0.172 \\ -0.086 & 0.733 \end{bmatrix} \begin{bmatrix} x_1(n) \\ x_2(n) \end{bmatrix} + \begin{bmatrix} 0.0172 \\ 0.0733 \end{bmatrix} e^{-0.15n}$$

or in longhand notation by

$$x_1(n+1) = 0.991x_1(n) + 0.172x_2(n) + 0.0172e^{-0.15n}$$

$$x_2(n+1) = -0.086x_1(n) + 0.733x_2(n) + 0.0733e^{-0.15n}$$

The plots in Figs. 5.1.2a and 5.1.2b compare the approximate numerical solution found from the following program to the analytical solution. Equation 3.1.13 of Chapter 3 gives the analytical solution. The program below produces Fig. 5.1.2b, where y_2 is the analytical solution and x_2 is the numerical solution.

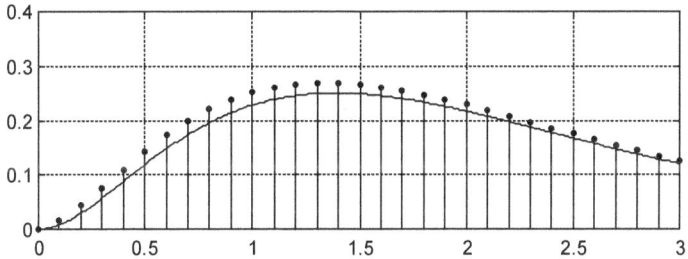

Fig. 5.1.2a. Plots of x_1.

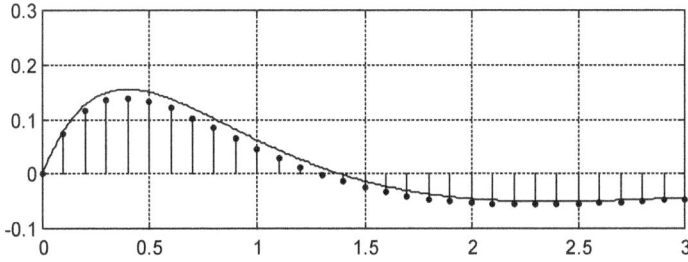

Fig. 5.1.2b. Plots of x_2.

5.9

```
clear
t3=linspace(0,3,301);
e=exp(-1.5*t3);
r=exp(-t3);
s=exp(-2*t3);
y1=-8*e+4*r+4*s;
y2=6*e-2*r-4*s;
subplot(2,1,1)
plot(t3,y2,'k')
grid on
hold on
t=linspace(0,3,31);
x20=0;
x10=0;
tn=0;
x1=0;
x2=0;
for i=1:31
    x(i)=x2;
    x1=0.991*x10+0.172*x20+0.0172*exp(-0.15*tn);
    x2=-0.086*x10+0.733*x20+0.0733*exp(-0.15*tn);
    tn=tn+1;
    x10=x1;
    x20=x2;
end
stem(t,x,'k.')
```

Type B Numerical solution

There is nothing new in this section. You've seen it all before. Simply change the type B equation to type A and proceed as above. We will work through an example for the sake of clarity.

The linear state equations for type B networks are of the form

$$\dot{x} = Ax + Bq + E\dot{q} \qquad (5.1.14)$$

As in objective 4.2, make the change of variable

$$z = x - Eq \qquad (5.1.15)$$

Substitute this into Eq. 5.1.14 to obtain the type A equation

$$\dot{z} = Az + (B + AE)q \qquad (5.1.16)$$

After finding the numerical solution for z, solve for x from Eq. 5.1.15.

Problem 5.1.6. Find the form of the numerical solution for the system in Problem 4.1.1, given by

$$\dot{i}_2 = -\tfrac{1}{2}i_2 + \tfrac{1}{2}\delta(t) - \tfrac{1}{2}e^{-t},\ t \geq 0 \qquad (A)$$

with initial condition $i_2(0^-) = 1$.

Solution: Identify the following terms:

$q(t) = $ _____ $A = $ _____

$E = $ _____ $B = $ _____

Now write Eq. (A) in the form given by Eq. 5.1.16.

$\dot{z}(t) = $ _____   (B)

with initial condition $z(0^-) = $ ____. (Obtain this initial condition from Eq. 5.1.15.) Note that since Eq. (B) satisfies the continuity theorem, $z(0^-) = z(0^+)$.) Now find ϕ and Δ.

Answers: $q(t) = e^{-t}u(t)$ $A = -\tfrac{1}{2}$

$$E = \tfrac{1}{2} \qquad\qquad B = 0$$

$$\dot{z}(t) = -\tfrac{1}{2}x(t) - \tfrac{1}{2}e^{-t},\ t \geq 0 \qquad (B)$$

$$z(0^-) = 1.$$

5.11

Self Test, Objective 5.1.

For your exam over this objective you will be given two problems, one type A and one type B. Therefore you should be able to solve the example problems in this objective.

Chapter 6

Transfer Function

Objectives: After completing this chapter you should be able to do the following:

6.1. Describe (define) transfer function.

6.2. Find the transfer function $H(s)$ for a given LTI circuit.

6.3. Find the steady-state response of an LTI circuit to an eternal exponential signal.

Rationale: The transfer function $H(s)$ relates the input to the output of an LTI system when the input is an exponential e^{st}. Therefore, the transfer function $H(s)$ is important if (1) it is easily found for most systems and (2) if most input signals can be expressed as the sum of exponential signals. Most LTI electrical networks and most signals meet both these conditions.

Another factor that makes the transfer function important is the ease with which it can be measured. This chapter concludes with a laboratory experiment in which the student is asked to measure $H(s)$ for four different systems.

Objective 6.1. Describe (define) transfer function.

6.1.1. Review of LTI Systems

This chapter begins our study of frequency analysis. The cornerstone of frequency analysis is the LTI properties. Frequency analysis applies the LTI properties directly to signals that are expressed as the sum of elementary signals, much as in Chapter 1. If the response to these elementary signals is known, we can find the response to the sum by the LTI properties. Here is a review.

Problem 6.1.1. An LTI system has the response $q(t)$ to the input $p(t)$ shown in Fig. 6.1.1. Find $y_1(t)$, the response to $x_1(t)$.

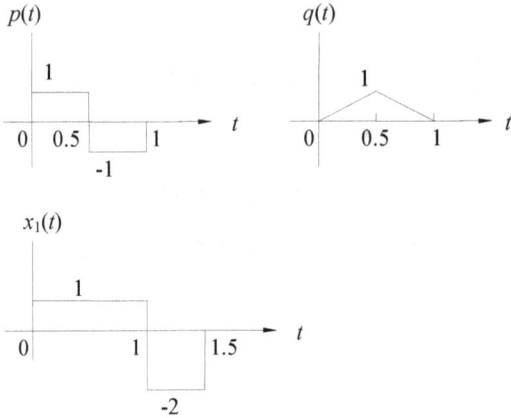

Fig. 6.1.1.

Solution: Fortunately $x_1(t)$ can be expressed as the sum of translated $p(t)$ terms as

$$x_1(t) = p(t) + 2p\left(t - \tfrac{1}{2}\right)$$

Apply the LTI properties to find $y_1(t)$.

$$y_1(t) = \underline{\quad} q(t) + \underline{\quad} q(t - \tfrac{1}{2})$$

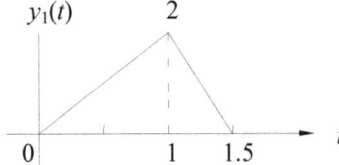

Fig. 6.1.2.

6.2

Answer: $y_1(t) = q(t) + 2q(t - \frac{1}{2})$ shown in Fig. 6.1.2.

Problem 6.1.2. For the system in problem 6.1.1, find the response to $x_2(t)$ shown in Fig. 6.1.3.

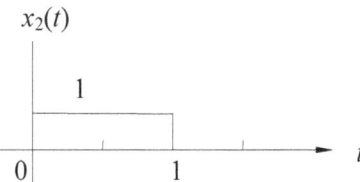

$x_2(t)$

1

0   1 t

Fig. 6.1.3.

Solution: This is a bit different, as you can appreciate if you work the problem first before looking at the answer. Express $x_2(t)$ as the sum of terms related to $p(t)$. To do this plot $p(t)$, $p(t - \frac{1}{2})$, and $p(t - 1)$. Then see what you have to do in order to obtain $x_2(t)$.

$x_2(t) = $ _____

Now apply the LTI properties to find the response $y_2(t)$.

$y_2(t) = $ _____

Answers:

$$x_2(t) = p(t) + \sum_{k=1}^{\infty} 2p(t - \frac{k}{2})$$

$$y_2(t) = q(t) + \sum_{k=1}^{\infty} 2q(t - \frac{k}{2})$$

6.3

Notes: a) A mathematician might object to this solution, but our objective here is to illustrate the technique, and this less-than-elegant example does so.

b) We were fortunate in both problems 6.1.1 and 6.1.2 in being able to express the input $x(t)$ as the sum of terms related to $p(t)$. But suppose the input $x(t)$ could not be expressed as the sum of $p(t)$ terms. (If $x(t)$ were a sinusoid, for example.) Then, using the methods we have discussed so far, it would be impossible to determine the response.

This leads to the question, is there an input signal with the following properties:

1. The system response to this input signal can easily be determined.

2. Most other input signals can be expressed as the sum of terms related to this input signal.

The answer is yes, there are several such signals. One is the unit impulse, to be used in connection with convolution, along with its cousin, the unit step. Another is the exponential signal such as $e^{j\omega t}$ or e^{st}. The use of these signals forms the basis of frequency analysis.

Definition of Transfer Function $H(s)$

It is a relatively simple matter to characterize an LTI system by the input-output pair

$$e^{st} - response\ to\ e^{st}$$

where s is a real or complex number called the frequency. We begin by considering an example of a system with e^{st} forcing function.

Problem 6.1.3. Find the response $v_2(t)$ if the input to the RC low-pass filter in Fig. 6.1.4 is an exponential e^{st}.

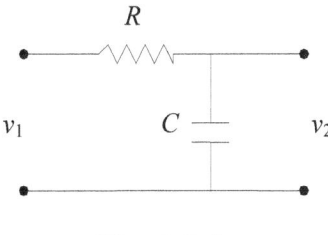

Fig. 6.1.4

Solution: The formulation procedure for type A networks gives the state equation

$$\frac{dv_2}{dt} = -\frac{v_2}{RC} + \frac{v_1}{RC}$$

Let us work quickly through the solution. With $v_1(t) = e^{st}$ the particular solution is

$$v_{2p}(t) = ke^{st} = \frac{(1/RC)}{s + (1/RC)}e^{st}$$

and the homogeneous solution is

$$v_{2h}(t) = ae^{-t/RC}$$

The sum of these two terms is the complete solution.

$$v_2(t) = \frac{(1/RC)}{s + (1/RC)}e^{st} + ae^{-t/RC}$$

where the constant a is determined from initial conditions.

With $H(s)$ defined by

$$H(s) = \frac{1/RC}{s + 1/RC}$$

the voltage $v_2(t)$ is composed of two parts. The forced or steady-state response is $H(s)e^{st}$ and the source free or transient response is $ae^{-t/RC}$. The term $H(s)$ is called the transfer function, system function, or filter characteristic of the network. Note that it is defined by

$$\frac{steady-state\ response\ to\ e^{st}}{e^{st}} = \frac{H(s)e^{st}}{e^{st}} = H(s)$$

This definition is valid so long as two conditions are met: (1) The system is asymptotically stable and (2) the system is physically realizable.

A system is asymptotically stable if its state always returns to the same value (usually zero) after the input is removed. A system is physically realizable if the output does not occur before the input is applied. For future reference here is a formal definition:

Definition 6.1.1. Transfer Function. For an LTI, asymptotically stable, physically realizable system, the function

$$H(s) = \frac{steady-state\ response\ to\ e^{st}}{e^{st}} \qquad (6.1.1)$$

is said to be the transfer function of the system.

Note: Another definition in terms of the impulse response $h(t)$ is given by

$$H(s) = \int_{-\infty}^{\infty} h(t)e^{-st}\,dt \qquad (6.1.2)$$

where it is necessary only that $h(t)$ be Laplace transformable. Equation 6.1.1 is equivalent to Eq. 6.1.2 under the conditions of asymptotic stability and physical realizability.

Self Test, Objective 6.1. Define transfer function.

Objective 6.2. Find the transfer function H(s) for given LTI circuits.

As stated earlier, it is a relatively simple matter to characterize an LTI system by the transfer function. Here is a procedure to use with electric circuits. This procedure can be justified by going the longer route of solving the system differential equation with e^{st} forcing function.

1. Replace each capacitor by the equivalent impedance $Z_C = 1/sC$.

2. Replace each inductor by the equivalent impedance $Z_L = sL$.

3. Resistors remain unchanged. That is, $Z_R = R$.

4. Find the ratio of response/input as in dc circuit theory. This ratio is $H(s)$.

For example, replace the circuit in Fig. 6.1.4 by the equivalent circuit in Fig. 6.2.1.

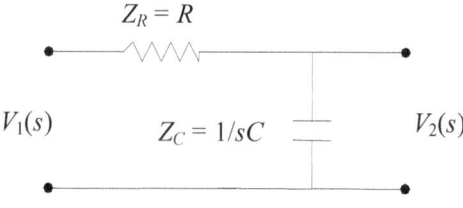

Fig. 6.2.1

The ratio $V_2(s)/V_1(s)$ is the transfer function, given by

$$H(s) = \frac{V_2(s)}{V_1(s)} = \frac{1/sC}{R + 1/sC} = \frac{1/RC}{s + 1/RC}$$

Problem 6.2.1. Find the transfer function $I_L(s)/V(s)$ for the circuit in Fig. 6.2.2.

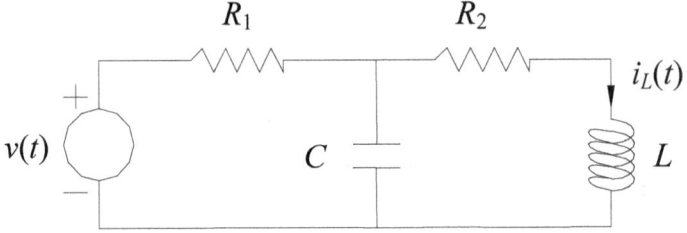

Fig. 6.2.2

Solution: Replace the elements by their impedance to obtain the circuit in Fig. 6.2.3.

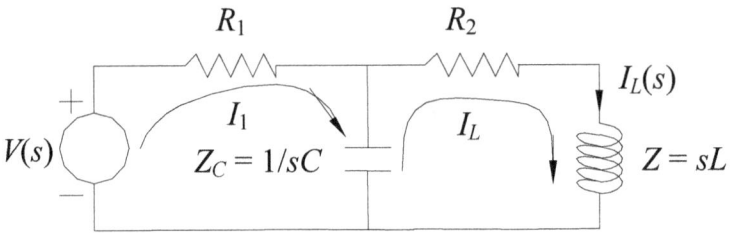

Fig. 6.2.3

The procedure for solution is now identical to that used in dc circuit theory. There are several equivalent methods available to us. Probably the most common method is to write loop equations, which give

$$V(s) = I_1\left(R_1 + \tfrac{1}{sC}\right) - \tfrac{I_L}{sC}$$

$$0 = -\tfrac{I_1}{sC} + I_L\left(R_2 + sL + \tfrac{1}{sC}\right)$$

Or

$$I_L(s) = \frac{\begin{vmatrix} (R_1 + \frac{1}{sC}) & V_s \\ -\frac{1}{sC} & 0 \end{vmatrix}}{\begin{vmatrix} (R_1 + \frac{1}{sC}) & -\frac{1}{sC} \\ -\frac{1}{sC} & (R_2 + sL + \frac{1}{sC}) \end{vmatrix}} = \frac{V(s)}{s^2 LCR_1 + s(CR_1 R_2 + L) + (R_1 + R_2)}$$

Therefore H(s) is given by

$$H(s) = \frac{I_L(s)}{V(s)} = \frac{1}{s^2 LCR_1 + s(CR_1 R_2 + L) + (R_1 + R_2)}$$

Note: The transfer function is a relation between the system input and output, where the input and output can be any measurable parameters. In problem 6.1.3 the transfer function has the units of admittance since the input excitation is a voltage and the response is a current. The point is that $H(s)$ can be the ratio of any current or voltage to any other current or voltage in the circuit.

Problem 6.2.2. A typical equivalent circuit for an electronic circuit is shown in Fig. 6.2.4. Find the voltage gain V_2/V_1.

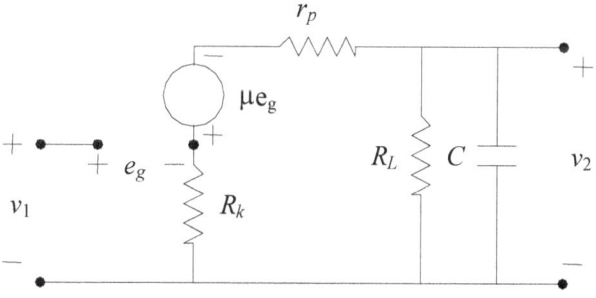

Fig. 6.2.4

Solution: First replace the parallel *RC* circuit by an equivalent impedance.

$$Z_L = \frac{R_L/sC}{R_L + \frac{1}{sC}} = \frac{R_L}{sCR_L + 1}$$

as shown in Fig. 6.2.5.

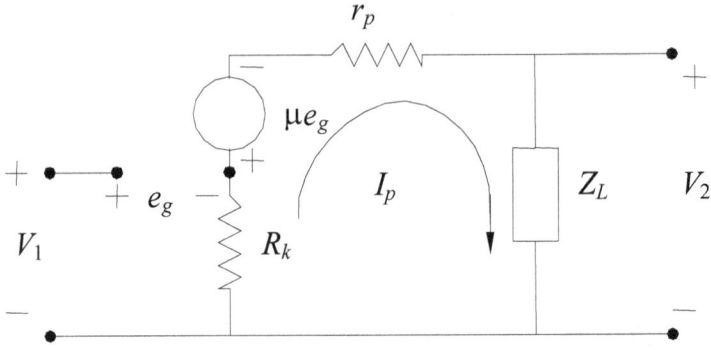

Fig. 6.2.5

Now write two loop equations.

$$V_1 - e_g + I_p R_k = 0$$

$$0 = I_p R_k + \mu e_g + I_p Z_L$$

Solve for e_g from the first equation and substitute into the second equation to obtain

$$0 = I_p \left[(\mu + 1) R_k + r_p + Z_L \right] + \mu V_1$$

With $I_p = V_2/Z_L$ this gives

$$0 = \frac{V_2}{Z_L} \left[(\mu + 1) R_k + r_p + Z_L \right] + \mu V_1$$

Or, finally,

$$H(s) = \frac{V_2}{V_1} = \frac{-\mu Z_L}{(\mu + 1)R_k + r_p + Z_L)}$$

$$= \frac{-\mu R_L}{s[CR_L R_k(\mu + 1) + CR_L r_p] + (\mu + 1)R_k + r_p + R_L}$$

Self Test, Objective 6.2. Find the transfer function $H(s)$ for the circuit in Fig. 6.2.6. The input is $v_1(t)$ and the output is $i_2(t)$.

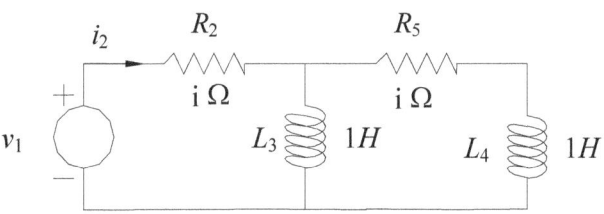

Fig. 6.2.6

Objective 6.3. Find the steady-state response of an LTI circuit to an eternal exponential signal e^{st}, $-\infty < t < \infty$.

Since we already know that the steady-state response to a signal e^{st} is $H(s)e^{st}$ for LTI systems, the purpose of this section is to gain some familiarity with the computation of this response.

Problem 6.3.1. Determine the steady-state response $v_2(t)$ to an input signal $v_1(t) = e^{(-1+j2)t}$ in Fig. 6.3.1.

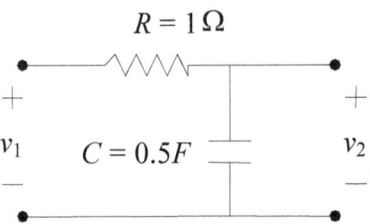

$$R = 1\,\Omega$$

$$C = 0.5F$$

Fig. 6.3.1

Solution: The transfer function, found earlier, is given by

$$H(s) = \frac{1/RC}{s + 1/RC}$$

Substitute the value $s = -1 + j2$ into this equation, along with the value of RC in Fig. 6.3.1 to obtain

$H(-1 + j2) = $ _____

Express this in polar form, then multiply by $v_1(t)$.

Steady-state part of $v_2(t)$ is $e^{(-1+j2)t}H(-1+j2) = $ _____

Answers to Problem 6.3.1.

$$H(-1 + j2) = \frac{2}{1 + j2} = 0.895e^{-j63.4}$$

The steady-state part of $v_2(t)$ is

$$v_2(t) = 0.895e^{[(-1+j2)t - j63.4°]}$$

Problem 6.3.2. Find the steady-state response of the circuit in Fig.6.3.2 to the input signal $v_1(t) = 10e^{(2+j1)t}$.

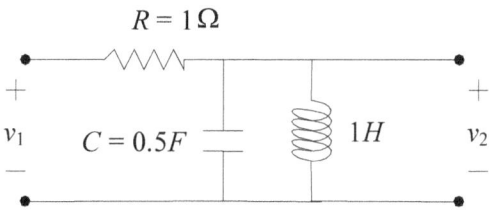

Fig. 6.3.2.

Solution: Compute the transfer function $H(s)$.

$\qquad H(s) = $ _____

Now express $H(2 + j1) = $ _____

The steady-state part of $v_2(t)$ is given by

$\qquad 10e^{(2+j1)t} H(2 + j1) = $ _____

Answers to Problem 6.3.2.

$$H(s) = \frac{s}{s^2 + s + 2}$$

$$H(2 + j1) = \frac{2 + j1}{7 + j5} = 0.258e^{j9°}$$

Steady-state part of $v_2(t)$ is

$$2.58e^{2t}e^{j(t+1°)}$$

Measuring H(s)

The transfer function for a stable LTI system can be measured easily and accurately. This is one reason the transfer function is important in engineering.

If the input is $e^{j\omega t}$ then, after the transients have died out, the response is $H(j\omega)e^{j\omega t}$. An exponential signal $e^{j\omega t}$ cannot be generated in the laboratory, but the sum of two such signals can be since

$$\cos \omega t = \frac{1}{2}e^{j\omega t} + \frac{1}{2}e^{-j\omega t}$$

$$\sin \omega t = \frac{1}{2j}e^{j\omega t} - \frac{1}{2j}e^{-j\omega t}$$

Imagine the following experiment in the laboratory. A sinusoidal signal generator supplies the input signal $v_1(t) = \cos \omega t$ to the RC circuit in Fig. 6.3.3. The output voltage $v_2(t)$ is measured across the capacitor by an oscilloscope or other recording device. If the amplitude of $v_1(t)$ is held constant as the frequency is varied, what happens to the output voltage $v_2(t)$?

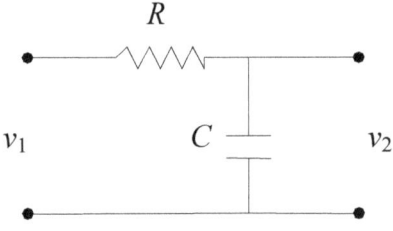

Fig. 6.3.3

At low frequencies (near dc) the capacitor acts like an open circuit and $v_2(t) \approx v_1(t)$. At high frequencies the capacitor acts like a short circuit and $v_2(t) \approx 0$. Thus the amplitude of $v_2(t)$ is reduced as frequency increases and the response curve of Fig. 6.3.4 will be plotted in the laboratory experiment.

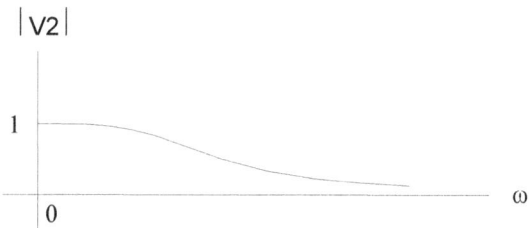

Fig. 11 4

The frequency of the output signal will always be the same as the frequency of the input signal. The phase angle will change, however. By this we mean that if $v_1(t) = \cos \omega t$ is the input signal then the output can be written in the form $v_2(t) = A_2 \cos(\omega t + \phi)$ where A_2 is the amplitude and ϕ is the phase angle. For example, if the input frequency is $\omega = 1/RC$ then the output is $v_2(t) = 0.707 \cos(\omega t - 45°)$. Thus in our laboratory experiment we can also plot the output phase angle as a function of frequency. Figure 6.3.5 would be the result.

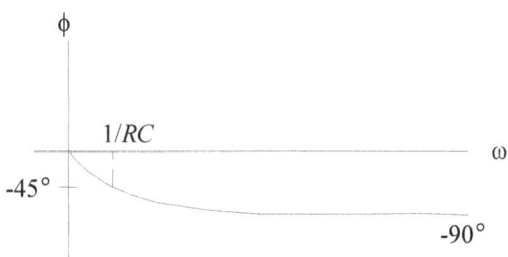

Fig. 6.3.5

Problem 6.3.3. If the steady state response to $e^{j\omega t}$ is $H(j\omega)e^{j\omega t}$, write down the steady state response to $\cos \omega t$.

Solution: Since $\cos \omega t = \frac{1}{2}e^{j\omega t} + \frac{1}{2}e^{-j\omega t}$, find the response to each exponential.

Steady state response to $\frac{1}{2}e^{j\omega t}$ = _____

Steady state response to $\frac{1}{2}e^{-j\omega t}$ = _____

So, steady state response to $\cos\omega t$ = _____

Answer:

Steady state response to $\frac{1}{2}e^{j\omega t} = \frac{1}{2}H(j\omega)e^{j\omega t}$.

Steady state response to $\frac{1}{2}e^{-j\omega t} = \frac{1}{2}H(-j\omega)e^{-j\omega t}$.

So, steady state response to $\cos\omega t$ is given by

$$\frac{1}{2}H(j\omega)e^{j\omega t} + \frac{1}{2}H(-j\omega)e^{-j\omega t} \qquad (6.3.1)$$

But this is not $H(j\omega)\cos\omega t$, as the unwary might guess. Refer to Figs. 6.3.4 and 6.3.5. There the response at the frequency $\omega = 1/RC$ was $0.707\cos(\omega t - 45°)$ which is not of the form $H(j\omega)\cos\omega t$. Equation 6.3.1 can be better expressed as follows.

First, express $H(j\omega)$ in polar form by

$$H(j\omega) = A(\omega)e^{j\theta(\omega)} \qquad (6.3.2)$$

Where $A(\omega)$ is the amplitude, and $\theta(\omega)$ is the angle. Now notice that $H(j\omega)$ is the complex conjugate of $H(-j\omega)$. This follows from the LTI properties of the system. Since $e^{j\omega t}$ is the conjugate of $e^{-j\omega t}$, then the response $H(j\omega)e^{j\omega t}$ is the conjugate of $H(-j\omega)e^{-j\omega t}$. Therefore $H(-j\omega)$ is given by

$$H(-j\omega) = A(\omega)e^{-j\theta(\omega)}$$

This allows to rewrite Eq. 6.3.1 as

$$\tfrac{1}{2}H(j\omega)e^{j\omega t} + \tfrac{1}{2}H(-j\omega)e^{-j\omega t}$$

$$= \tfrac{1}{2}A(\omega)e^{j[\omega t + \theta(\omega)]} + \tfrac{1}{2}A(\omega)e^{-j[\omega t + \theta(\omega)]}$$

$$= A(\omega)\cos[\omega t + \theta(\omega)] \qquad (6.3.3)$$

This now conforms to our experimentally derived results in Figs. 6.3.4 and 6.3.5.

Note: There is a difference between amplitude and magnitude. Amplitude can have both positive and negative real values. Magnitude is restricted to only positive real values. Our work is much easier if we use amplitude instead of magnitude.

Problem 6.3.4. Determine the steady state response to a sinusoidal signal $v_1(t) = 2\sin 2\pi t$, $-\infty < t < \infty$, in Fig. 6.3.6.

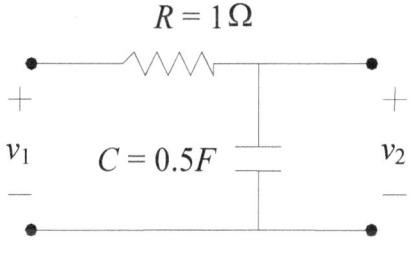

$$R = 1\,\Omega$$

$$v_1 \qquad C = 0.5F \qquad v_2$$

Fig. 6.3.6.

Solution: Use Euler's formula to express $v_1(t)$ as the sum of two exponential signals.

$$v_1(t) = \underline{\hspace{3cm}}$$

We have previously found H(s) for this circuit. It is given by

$$H(s) = \underline{\hspace{3cm}}$$

6.17

The frequency of the input signal is $s_1 = j2\pi$ and $s_2 = -j2\pi$. Calculate $H(s)$ for these two values of frequency.

$$H(j2\pi) = \underline{\hspace{3cm}}$$

$$H(-j2\pi) = \underline{\hspace{3cm}}$$

Now calculate the response of the network to each exponential.

$$\frac{1}{j}e^{j2\pi t}H(j2\pi) = \underline{\hspace{3cm}}$$

$$-\frac{1}{j}e^{-j2\pi t}H(-j2\pi) = \underline{\hspace{3cm}}$$

Finally, sum these two responses to obtain $v_2(t)$. Express $v_2(t)$ as a sinusoid.

$$v_2(t) = \underline{\hspace{3cm}}$$

Answers:

$$v_1(t) = \frac{1}{j}e^{j2\pi t} - \frac{1}{j}e^{-j2\pi t}$$

$$H(s) = \frac{2}{s+2}$$

$$H(j2\pi) = \frac{2}{2+j2\pi} = 0.303e^{-j72.4\circ}$$

$$H(-j2\pi) = \frac{2}{2-j2\pi} = 0.303e^{j72.4\circ}$$

$$\frac{1}{j}e^{j2\pi t}H(j2\pi t) = \frac{0.303}{j}e^{j(2\pi t-72.4°)}$$

$$-\frac{1}{j}e^{-j2\pi t}H(-j2\pi t) = -\frac{0.303}{j}e^{-j(2\pi t-72.4°)}$$

$$v_2(t) = 0.607\sin(2\pi t - 72.4°)$$

Problem 6.3.5. Find the steady state response of the circuit in Fig. 6.3.7 to the signal $v_1(t) = 1 + \cos(2\pi t)$.

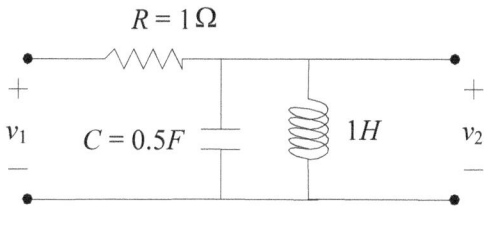

Fig. 6.3.7.

Answer: $v_2(t) = 0.165\cos(2\pi t - 80.5°)$.

Self Test, Objective 6.3.

1. Find the transfer function $H(s)$ for the circuit in Fig. 6.3.8. The input is $v_1(t)$ and the output is $v_2(t)$.

2. Find the steady state response if the input signal is given by

a) $e^{(1-j2)t}, \quad -\infty < t < \infty$

b) $e^{(-2+j3)t}, \quad -\infty < t < \infty$

c) $\sin t, \quad -\infty < t < \infty$

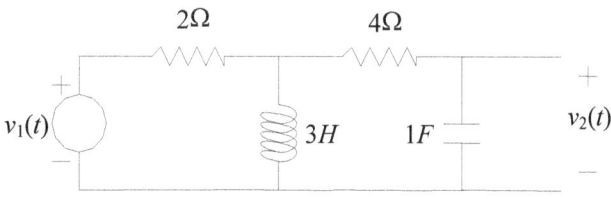

Fig. 6.3.8

6.19

Laboratory Experiment: There are four "black boxes" that your instructor will provide for you. Each box has two input terminals and two output terminals.

1. Measure and plot $H(j\omega)$ for each box.
2. Try to guess what the circuit is like in each box.
3. Open the box, copy the circuit, and calculate the transfer function $H(j\omega)$.

Self Test Answers:

Objective 6.1: See definition 6.1.1.

Objective 6.2: $H(s) = \dfrac{2s+1}{s^2 + 3s + 1}$

Objective 6.3:

1. $H(s) = \dfrac{3s}{18s^2 + 11s + 2}$

2. a) $0.065e^t e^{-j(2t-50°)}$

   b) $0.0505e^{-2t}e^{\wedge}j(3t - 55.4°)$

   c) $0.154\sin(t - 55°)$

Chapter 7

Fourier Series

Objectives: After completing this chapter you should be able to do the following:

7.1   Describe (define) vector.
7.2   Calculate the inner product of two given waveforms.
7.3   Determine (select) which functions have a Fourier series.
7.4   Calculate and plot the Fourier series for a given waveform.
7.5   Calculate and plot the waveform for a given Fourier series.

Rationale:

Aside from the fact that frequency analysis is one of three ways to analyze LIT systems, the use of the Fourier series provides valuable insight into system behavior. Historically, one of the first electrical engineering applications of frequency analysis was to radio and telephone transmission systems. The operation of these systems depends on their ability to select frequency bands in the transmission and reception circuits. Hence, the frequency content of signals is all-important. This chapter discusses the mechanics of decomposing a signal into its frequency components, and the next chapter considers the effects of systems on these frequency components.

Objective 7.1. Describe (define) vector.

In Chapter 6 we said that the signal e^{st} has two desirable properties. First, the response of most LTI systems to this input signal could easily be determined, and second, most other input signals can be expressed as the sum of terms related to this signal. In Chapter 6 we demonstrated the first property. Before considering the second property you should have a clear understanding of geometry, because signals are vectors, and signal processing is applied geometry.

An arbitrary signal can be expressed as the sum of exponential signals because $e^{jn\omega t}$ is orthogonal to $e^{jm\omega t}$ if $n \neq m$. But what is meant by orthogonality of signals? Many engineers are at home when discussing orthogonality of geometric vectors, but not signals.

What do you think a vector is? If you are standing on the corner and someone asks, "What is a vector?" What do you say?

I hate to tell you this, but if you said that a vector is a directed magnitude, then you are wrong.

The definition of vector probably evolved, as did most definitions, through a long process of trial and error until a satisfactory definition could be agreed upon. Our present definition is a list of the important properties of ordinary geometric vectors. Then any set of objects that has these properties is called a set of vectors. It turns out that a multitude of sets have these properties. Thus the following defines a *vector space*, not a vector. A vector space is a set of objects, and each member of this set is a vector.

Definition 7.1. A *vector space* is a set $V = \{v_i\}$ together with a field of scalars $A = \{a_i\}$ that has the following two operations and seven properties.

a) We can add two vectors together and obtain a third vector. Thus there is a mechanism for combining two vectors to obtain a third.

b) We can multiply a vector by a scalar and obtain another vector. This mechanism combines a scalar with a vector to obtain another vector.

Using these two operations, vector addition and scalar multiplication, the following properties must hold for all $v_i \in V$ and all $a_i \in A$.

1. $v_1 + v_2 = v_2 + v_1$

2. $(v_1 + v_2) + v_3 = v_1 + (v_2 + v_3)$

3. $a_i(v_1 + v_2) = a_i v_1 + a_i v_2$

4. $(a_1 + a_2)v_i = a_1 v_i + a_2 v_i$

---

5. $a_1(a_2 v_i) = (a_1 a_2)v_i$

6. $1 \cdot v_i = v_i$

7. There exists a unique vector v_0, called the zero vector, such that for all vectors v_i we have $0\ v_i = v_0$, where 0 is the number zero.

Note: The only scalars in this book are ordinary real or complex numbers.

Self Test, Objective 7.1.

Which of the following sets of objects satisfy Definition 7.1, and therefore can be called a vector space?

1. All 4X1 matrices such as

$$X = \begin{bmatrix} x_1 \\ x_2 \\ x_3 \\ x_4 \end{bmatrix} \quad Y = \begin{bmatrix} y_1 \\ y_2 \\ y_3 \\ y_4 \end{bmatrix} \quad Z = \begin{bmatrix} z_1 \\ z_2 \\ z_3 \\ z_4 \end{bmatrix}$$

2. All 3X2 matrices such as

$$X = \begin{bmatrix} x_{11} & x_{12} \\ x_{21} & x_{22} \\ x_{31} & x_{32} \end{bmatrix} \quad Y = \begin{bmatrix} y_{11} & y_{12} \\ y_{21} & y_{22} \\ y_{31} & y_{32} \end{bmatrix} \quad Z = \begin{bmatrix} z_{11} & z_{12} \\ z_{21} & z_{22} \\ z_{31} & z_{32} \end{bmatrix}$$

3. All matrices of any finite dimension.

4. Real numbers such as $X = 21.5$, $Y = -3$, $Z = 0.12$.

5. Voltage waveforms that can be generated in the laboratory.

Objective 7.2. Calculate the inner product of two given waveforms.

If, in addition to the above seven properties, a dot product (also called inner product) is defined for the vectors, the set is called an inner product space. Ordinary geometric vectors with the usual definition of dot product form an inner product space.

Now consider ordinary signals that can be generated in the laboratory as functions of time. These functions $v_1(t)$, $v_2(t)$, $v_3(t)$, ... satisfy the seven properties listed above; hence they are vectors. Now define the inner product $\langle v_1 | v_2 \rangle$ as

$$\langle v_1 | v_2 \rangle = \int_{t_1}^{t_2} v_1(t) v_2(t) dt \qquad (7.2.1)$$

Then these signals form an inner product space.

Orthogonality: Two signals are said to be orthogonal over the interval $t_1 < t < t_2$ if

$$\langle v_1 | v_2 \rangle = 0 \qquad (7.2.2)$$

Notes: a) The interval $t_1 < t < t_2$ is important. It is obvious but worth stating that two functions may be orthogonal over one interval (t_1, t_2) but not over another interval (t_3, t_4).

b) Nothing is said about "right angles" in this definition of orthogonality. Geometric vectors can be at right angles, but this concept does not apply to waveforms. Remember, we have already scolded you for thinking of vectors as pointy things with direction and magnitude.

For real signals such as those generated in the laboratory Eq. 7.2.1 is a good definition of inner product. The Fourier series deals with complex-valued signals $e^{j\omega t}$, so our definition must be expanded to the following:

7.4

Definition 7.2.1. Inner Product of Time Functions. The inner product (dot product) of two time functions is given by

$$\langle v_1|v_2 \rangle = \int_{t_1}^{t_2} v_1(t)v_2^*(t)dt \qquad (7.2.3)$$

where $v^*(t)$ is the complex conjugate of $v(t)$.

Notes: a) If the two time functions are real, Eq. 7.2.3 reduces to Eq. 7.2.1.

b) The reason for the conjugate in Eq. 7.2.3 is so that the inner product will be a real number.

c) Think of the inner product as a black box with two inputs and one output. The two inputs are vectors and the output is a scalar.

Problem 7.2.1. Find the inner product $\langle v_1|v_2 \rangle$ for each set of waveforms in Fig. 7.2.1.

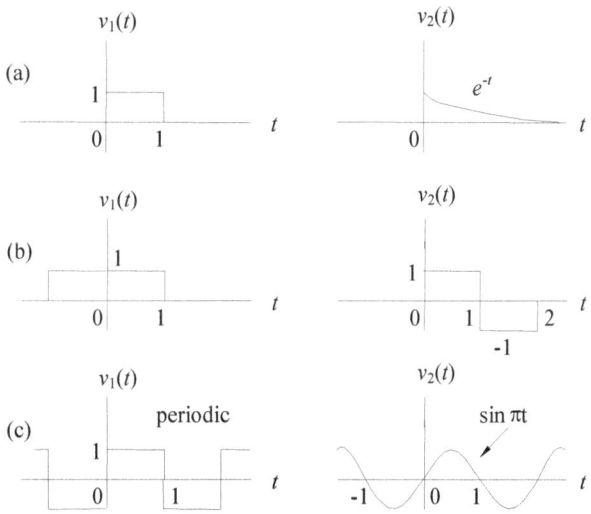

Fig. 7.2.1

Solution: Hopefully you ran into a problem. You cannot use the same inner product in each part. For the first two parts (a) and (b) you can use the following:

$$\langle v_1 | v_2 \rangle = \int\limits_{-\infty}^{\infty} v_1(t) v_2^*(t) dt$$

But infinite limits on the integral do not work in part c. Finite limits must be used, say

$$\langle v_1 | v_2 \rangle = \int\limits_{0}^{2} v_1(t) v_2^*(t) dt$$

Using these limits you should obtain the following, a) 0.633, b) 1.0, c) $4/\pi$.

The object of this problem is to demonstrate that there is no universally applicable inner product that can be used with all time functions. Different situations use different inner products.

There are many different types of vectors, but this chapter deals only with time functions. It is important to note that the only type of vector that has the property of "direction" is a geometric vector. As we noted earlier, right angles apply only to geometric vectors. Orthogonality means only that the inner product is zero. It implies nothing about right angles except for geometric vectors.

Self Test, Objective 7.2. Calculate the inner product of the two waveforms in Fig. 7.2.2.

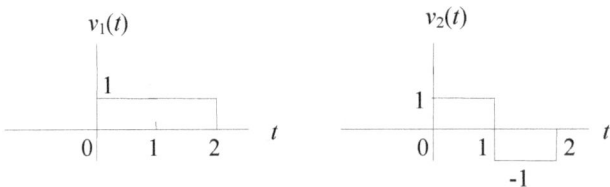

Fig. 7.2.2.

7.6

Objective 7.3. Determine (select) which functions have a Fourier series.

7.3.1. The Fourier Series as an Operator

An operator is a function whose domain and range are both sets of functions. From another viewpoint, an operator is a black box whose input and output are both vectors, as shown in Fig. 7.3.1. "Transform" is another name for the same thing. The Fourier series, Fourier transform, and Laplace transform are operators, that is, a black box with vectors at both the input and output. (Note that our mathematical model for a system is an operator.)

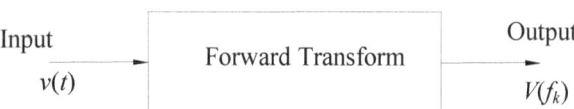

Fig. 7.3.1.

The Fourier series operator is the relationship between $v(t)$ and V_k or $V(f_k)$ in the following equation:

$$V_k = V(f_k) = \frac{1}{T}\int_{t_1}^{t_1+T} v(t)e^{-j\omega_k t}\,dt \qquad (7.3.1)$$

In Fig. 7.3.1 a particular function $v(t)$ is selected as the input to the black box. In Eq. 7.3.1 a particular function $v(t)$ is selected and supplied to the integral. Same thing.

In Fig. 7.3.1 the output is the function V_k or $V(f_k)$. In Eq. 7.3.1 the result of the integration and division by T is the function V_k or $V(f_k)$. Same thing.

Now v and V are themselves functions. The input v is a function of the continuous parameter t (by this we mean that values of t are in the domain of v), and V can be thought of as a function of either the index k or the discrete parameter f_k, in which case we write $V(f_k)$. Since we are concerned with signals as functions of time, we will identify t as time. The parameter f_k represents discrete frequency.

Note: The parameter t can represent any physical quantity or attribute. For example, one application has identified t as space and k as wave number. The French mathematician Fourier first used the series in his study of heat – an application that is physically different from ours.

All the transforms in this book have an inverse. That is, if we change things around and supply V_k to another black box labeled "Inverse Transform" the output will be $v(t)$. Furthermore, this $v(t)$ will equal the original $v(t)$. This inverse operation is

$$v(t) = \sum_{k=-\infty}^{\infty} V_k e^{j2\pi f_k t} \tag{7.3.2}$$

Again, this is an operator; it is illustrated in Fig. 7.3.2. The input is a function $V(f_k)$ = V_k, and the output is a function $v(t)$.

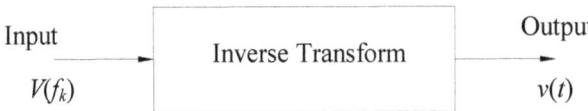

Input   Inverse Transform   Output

$V(f_k)$ $v(t)$

Fig. 7.3.2.

7.3.2. Which functions have a Fourier series?

Look at Eq. 7.3.1. If the magnitude of V_k is to be finite for each k, then it must be true that

$$\left| \int_{t_1}^{t_1+T} v(t) e^{-2\pi f_k t} dt \right| < \infty$$

Since the magnitude of $e^{j\theta}$ is unity for any θ then it will suffice to have

$$\int_{t_1}^{t_1+T} |v(t)| \, dt < 0 \qquad\qquad (7.3.3)$$

This is known as the weak Dirichlet (pronounced Der-Clay) condition. If, in addition to satisfying the weak Dirichlet condition, the function $v(t)$ is finite and has a finite number of maxima and minima in the interval t_1 to t_1+T, then it satisfies the strong Dirichlet conditions. We will take these three conditions as our criteria for possessing a Fourier series. They are:

1. $v(t)$ satisfies Eq. 7.3.3.
2. $v(t)$ is finite.
3. $v(t)$ has a finite number of maxima and minima.

Notes: a) Any function that can be generated in the laboratory has a Fourier series.

b) The strong Dirichlet conditions are only one of several sets of conditions that could be used. The problem of finding both necessary and sufficient conditions has not been solved.

c) Notice that it is not necessary for the function $v(t)$ to be periodic before it has a Fourier series, as you may assume from previous experience.

Self Test, Objective 7.3.

1. Describe one set of conditions that a waveform must meet before it has a Fourier series.

2. Which of the following waveforms has a Fourier series?

     a) $f(t) = \cos \omega t, \quad -\infty < t < \infty$

     b) $f(t) = \sin \omega t, \quad -\infty < t < \infty$

     c) $f(t)$ is a periodic square wave. (Fig. 7.3.3c.)

     d) $f(t)$ is a single square pulse. (Fig. 7.3.3d.)

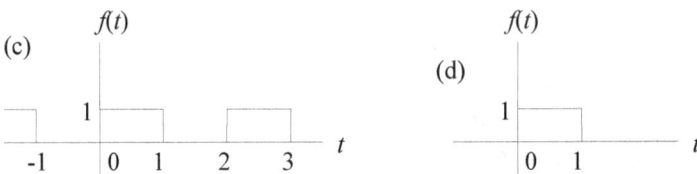

Fig. 7.3.3.

Objective 7.4. Calculate and plot the Fourier series for a given waveform.

7.4.1. Orthogonality Property

The exponential functions in the Fourier series are orthogonal to each other. Consider Eq. 7.3.2, rewritten here for convenience:

$$v(t) = \sum_{k=-\infty}^{\infty} V_k e^{j2\pi f_k t} \qquad (7.3.2)$$

With $f_k = kf_1$ and $\omega = 2\pi f$ this can be written as

$$v(t) = \cdots V_{-1}e^{-j\omega_1 t} + V_0 + V_1 e^{j\omega_1 t} + V_2 e^{j\omega_2 t} + V_3 e^{j\omega_3 t} + \cdots \qquad (7.4.1)$$

The exponential functions $g_k(t) = e^{j\omega_k t}, \cdots -2, -1, 0, 1, 2, \cdots$ are orthogonal over the interval $t_1 < t < t_1 + T$, where $T = 1/f_1$. This is easily seen by

$$\langle g_k | g_i \rangle = \int_{t_1}^{t_1+T} e^{jk\omega_1 t} e^{ji\omega_1 t} dt = \int_{t_1}^{t_1+T} e^{j(k-i)\omega_1 t} dt = \begin{cases} 0, & k \neq i \\ T, & k = i \end{cases}$$

This allows us to easily evaluate the coefficients in Eq. 7.3.2. For example, calculate V_2 as follows: Multiply both sides of Eq. 7.4.1 by $e^{-j2\omega_1 t}$ to get

$$v(t)e^{-j2\omega_1 t} = \cdots V_{-1}e^{-j3\omega_1 t} + V_0 e^{-j2\omega_1 t} + V_1 e^{-j\omega_1 t} + V_2 + V_3 e^{j\omega_1 t} + \cdots$$

Notice that every term contains an exponential except V_2. Now integrate both sides over any interval of length T to obtain

$$\int_{t_1}^{t_1+T} v(t)e^{-j2\omega_1 t}dt = \cdots + 0 + 0 + 0 + V_2 T + 0 + \cdots$$

Solving for V_2 gives

$$V_2 = \frac{1}{T}\int_{t_1}^{t_1+T} v(t)e^{-j2\omega_1 t}dt$$

Or, for general k,

$$V_k = \frac{1}{T}\int_{t_1}^{t_1+T} v(t)e^{-jk\omega_1 t}dt$$

This is Eq. 7.3.1.

Note: This shows that the orthogonality of the vectors $e^{-jk\omega_1 t}$ may be used to evaluate the coefficients V_k in the Fourier series. Furthermore, if a function $v(t)$ is written as a series, Eq. 7.3.2, this orthogonality property leads naturally to Eq. 7.3.1, the forward operation on $v(t)$.

7.4.2. Expansion of an Arbitrary Signal $v(t)$

Let us choose some interval of length T and expand an arbitrary function $v(t)$ in terms of the complex exponentials $e^{j\omega_k t}$. Then, so long as the Dirichlet conditions are satisfied, the coefficients V_k are found by inserting the equation for $v(t)$ between t_1 and t_1+T in the operator, Eq. 7.3.1. See Fig. 7.4.1.

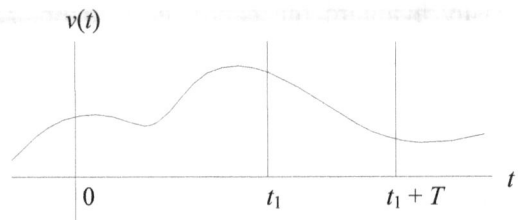

Fig. 7.4.1. An arbitrary function $v(t)$ to be expanded over the interval (t_1, t_1+T).

Having found the set of Fourier coefficients $\{V_k\}$ we may use Eq. 7.3.2 to obtain the time function $v(t)$ *in the interval* (t_1, t_1+T). The function $v(t)$ will be given by

$$v(t) = \sum_{k=-\infty}^{\infty} V_k e^{j2\pi f_k t} \tag{7.3.2}$$

Unfortunately, the set of numbers $\{V_k\}$ tells us nothing about the function $v(t)$ outside the interval (t_1, t_1+T). Thus we have lost something in transforming to the frequency domain $\{V_k\}$ and back to the time domain $v(t)$.

From another viewpoint, consider Fig. 7.4.2. The function $w(t)$ is identical to $v(t)$ in the interval (t_1, t_1+T), but is different outside this interval. If the coefficients $\{W_k\}$ corresponding to $w(t)$ in the interval (t_1, t_1+T) are found, they will be identical to $\{V_k\}$. The equations for $v(t)$ and $w(t)$ are identical in the interval (t_1, t_1+T), and this is the function used to calculate the coefficients in Eq. 7.3.1.

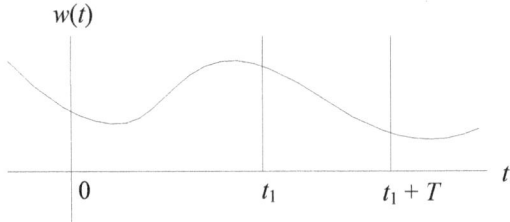

Fig. 7.4.2. The function $w(t)$ identical to $v(t)$ in the interval (t_1, t_1+T), but different elsewhere.

We conclude that the set of coefficients $\{V_k\}$ is sufficient to specify the time function $v(t)$ in the interval of expansion, but supplies no information about $v(t)$ outside this interval.

7.4.3. Fourier Series Expansion of Periodic Time functions

Consider a periodic time function $v(t)$ with period T. Since the set of coefficients $\{V_k\}$ completely specifies this time function for one period, the periodicity of the function specifies the function for all time. Therefore we need to know three things in order to completely specify a periodic time function by its Fourier series coefficients.

1. The time function is periodic with period T.

2. The period of expansion (t_1, t_1+T).

3. The set of coefficients $\{V_k\}$.

Problem 7.4.1. Find the Fourier coefficients for the periodic square wave in Fig. 7.4.3.

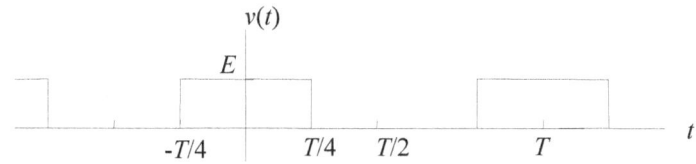

Fig. 7.4.3. Periodic square wave.

Solution: We may choose any interval of length T. The most convenient interval is $(-T/2, T/2)$. Equation 7.3.1 gives

$$V_k = \frac{1}{T} \int_{-T/2}^{T/2} v(t)e^{-j\omega_k t}\,dt = \frac{1}{T} \int_{-T/4}^{T/4} Ee^{-j\omega_k t}\,dt = E\left(\frac{e^{\frac{j\omega_k T}{4}} - e^{-\frac{j\omega_k T}{4}}}{j\omega_k T}\right)$$

Now $\omega_k = k\omega_1$ and ω_1 is the fundamental frequency related to T by $f_1 = 1/T$. Therefore $\omega_k T = k\omega_1 T = k2\pi$ and

$$V_k = E\left(\frac{e^{\left(\frac{jk2\pi}{4}\right)} - e^{-\left(\frac{jk2\pi}{4}\right)}}{j2\pi k}\right) = \frac{E}{2}\left(\frac{\sin k\pi/2}{k\pi/2}\right)$$

Figure 7.4.4 shows a picture of the coefficients V_k versus frequency. The continuous curve is the envelope

$$V(f) = \frac{E}{2}\left(\frac{\sin \omega T/4}{\omega T/4}\right)$$

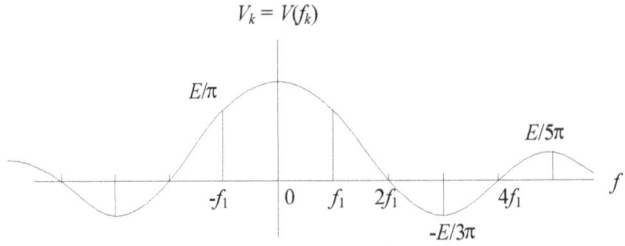

Fig. 7.4.4. The coefficients V_k versus frequency.

Notes: a) $V(f)$ is derived from Eq. 7.4.2 with $\omega = \omega_k$. The values of the coefficients V_k are given by the values of $V(f)$ evaluated at $V(f) = V(f_k) = V_k$. Thus the coefficients V_k are functions of frequency. Alternatively, think of the coefficients V_k simply as numbers, and if you know the fundamental period T, you can reconstruct the original periodic time function from these numbers.

7.14

b) Since the functions $e^{j\omega_k t}$ are periodic, the expansion given in Eq. 7.3.2 always yields a periodic function with period $T = 2\pi / \omega_1$. The expansion of the functions shown in Figs. 7.4.1 and 7.4.2, for example, yields periodic functions with period T.

c) In this example each coefficient is a real (not complex) number. This is a special case. In general, the numbers $\{V_k\}$ are complex. Therefore a graph of the coefficients versus frequency will usually require either a three-dimensional diagram or two separate graphs. In the next problem the coefficients are complex numbers.

Problem 7.4.2. Shift the periodic square wave in Fig. 7.4.3 to the right by $T/4$ to produce $g(t)$, as shown in Fig. 7.4.5. Find the coefficients $\{G_k\}$.

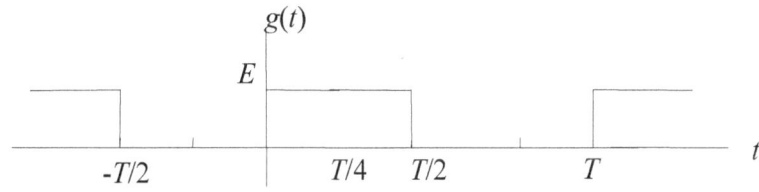

Fig. 7.4.5.

Solution:

$$G_k = \frac{1}{T} \int_0^{T/2} E e^{-j\omega_k t} dt = E e^{-\frac{j\omega_k T}{4}} \left(\frac{e^{\frac{j\omega_k T}{4}} - e^{-\frac{j\omega_k T}{4}}}{j\omega_k T} \right)$$

Compare this to Eq. 7.4.2 to see that G_k is related to V_k by

$$G_k = V_k e^{-\frac{j\omega_k T}{4}} = V_k e^{-\frac{jk\pi}{2}}$$

Figure 7.4.6 shows a plot of amplitude and phase for this function.

7.15

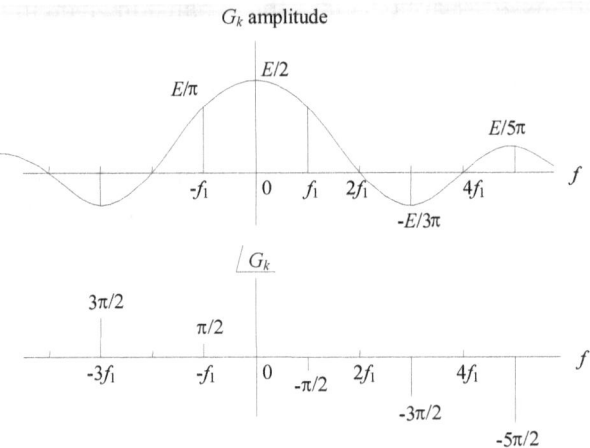

Fig. 7.4.6. Amplitude and phase of G_k.

Notes: a) There is a big difference between magnitude and amplitude. The magnitude of a number is always real and positive. Amplitude is also real but it can be negative. The amplitude plot in Fig. 7.4.6 contains negative values. The corresponding magnitude plot would contain only positive values, and the phase plot would be different.

b) Each coefficient G_k is a complex number. This means that G_k is not a single number, but is specified by two numbers (an ordered pair). For instance G_1 is given by its amplitude and phase as $(E/\pi, -\pi/2)$, which is conventionally written as

$$G_1 = \frac{E}{2}e^{-\frac{j\pi}{2}}$$

The magnitude and phase represent a complex number in polar form. Real and imaginary parts are used to represent this number in rectangular coordinates by $(0, -E/\pi)$ or, more conventionally,

$$G_1 = 0 - j\frac{E}{\pi}$$

Problem 7.4.3. The voltage was on half the period and off half the period in the previous two problems. The more general case has arbitrary pulse width Δ as shown in Fig. 7.4.7. Find the Fourier coefficients W_k.

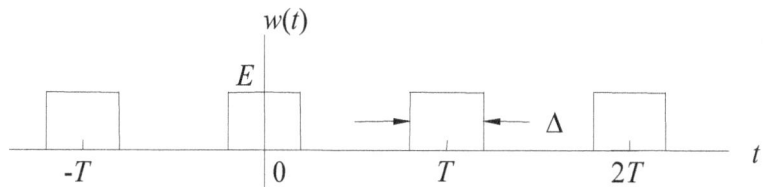

Fig. 7.4.7 Periodic square wave with arbitrary pulse width.

Solution:
$$W_k = \frac{1}{T}\int_{-\Delta/2}^{\Delta/2} E e^{-j\omega_k t}\,dt = \frac{\Delta E}{2}\frac{\sin k\pi\Delta/T}{k\pi\Delta/T} \tag{7.4.3}$$

Figure 7.4.8 shows the graph of W_k.

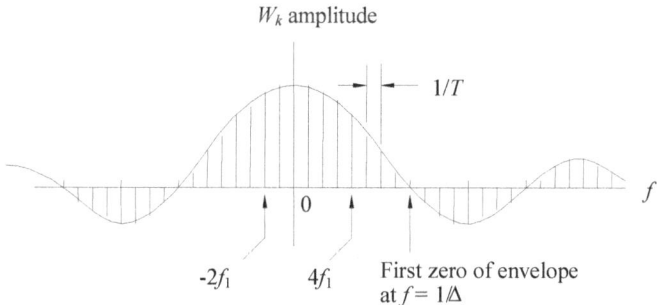

Fig. 7.4.8. The coefficients W_k.

Notes: a) The spacing between lines is the same as before. This spacing depends on T, the period of the time function. The $\sin(x)/x$ envelope depends on Δ, the pulse width. This $\sin(x)/x$ envelope is characteristic of periodic square pulses. Different pulse shapes result in different envelope shapes.

7.17

b) The coefficients are real here, just as they were in Problem 7.4.1. This is because the time function is even. An important property of Fourier series is that even time functions result in purely real coefficients and odd time functions result in purely imaginary coefficients. Other properties are discussed in Chapter 12.

Problem 7.4.4. Represent the function e^t (Fig. 7.4.9) over the interval $(0, 1)$ by the exponential Fourier series.

Solution: Here $T = 1$ and $\omega_1 = 2\pi/T = 2\pi$. Therefore

$$F_n = \frac{1}{T}\int_0^T v(t)e^{-jn\omega_1 t}dt = \int_0^1 e^t e^{-jn\omega_1 t}dt = \int_0^1 e^{(1-jn2\pi)t}dt = \frac{e^{1-jn2\pi} - 1}{1 - jn2\pi}$$

This gives an infinite number of terms. Figure 7.4.10 plots the first few.

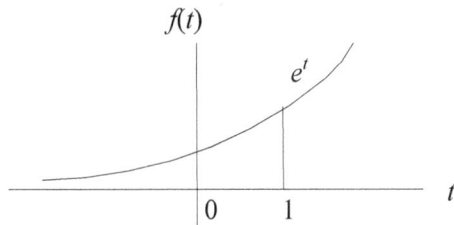

Fig. 7.4.9. The exponential e^t.

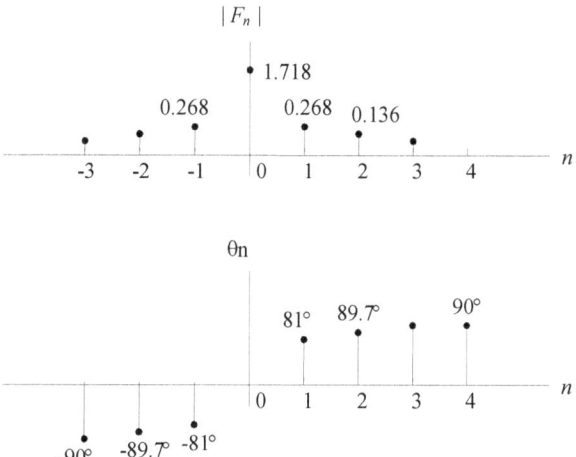

Fig. 7.4.10. The first few terms of F_n.

Self Test, Objective 7.4.

Calculate and plot the Fourier series for each waveform in Fig. 7.4.11.

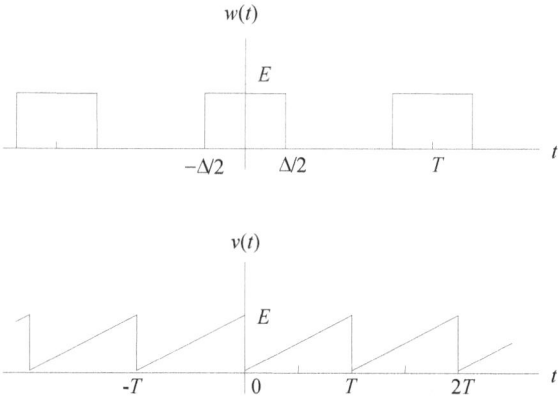

Fig. 7.4.11.

7.19

Objective 7.5. Calculate and plot the waveform for a given Fourier series.

Here are two problems to illustrate how to find the time function corresponding to Fourier coefficients. In each case we simply apply the definition, given by

$$v(t) = \sum_{k=-\infty}^{\infty} V_k e^{j2\pi f_k t}$$

Problem 7.5.1. The amplitude and phase spectrum of the periodic signal $v(t)$ are shown in Fig. 7.5.1. Write an analytic expression for $v(t)$ in terms of trigonometric functions.

Solution: The function $v(t)$ is given by

$$v(t) = \sum_{k=-\infty}^{\infty} V_k e^{j2\pi f_k t}, \quad t_1 < t < t_1 + T$$

where $\omega_1 = 2\pi f_1 = 2\pi(10)$. From Fig. 7.5.1 the coefficients are given by

$$V_{-2} = \tfrac{1}{4}e^{j90°}, \quad V_{-1} = \tfrac{1}{2}e^{j45°}$$

$$V_0 = 1, \quad V_1 = \tfrac{1}{2}e^{-j45°}, \quad V_2 = \tfrac{1}{4}e^{-j90°}$$

Therefore $v(t)$ is given by

$$v(t) = \tfrac{1}{4}e^{j90°}e^{-j2\pi(20)t} + \tfrac{1}{2}e^{j45°}e^{-j2\pi(10)t} + 1$$
$$+ \tfrac{1}{2}e^{-j45°}e^{j2\pi(10)t} + \tfrac{1}{4}e^{-j90°}e^{j2\pi(20)t}$$

Apply Euler's formula to obtain

$$v(t) = 1 + \cos(2\pi(10)t - 45°) + \tfrac{1}{2}\cos(2\pi(20)t - 90°)$$

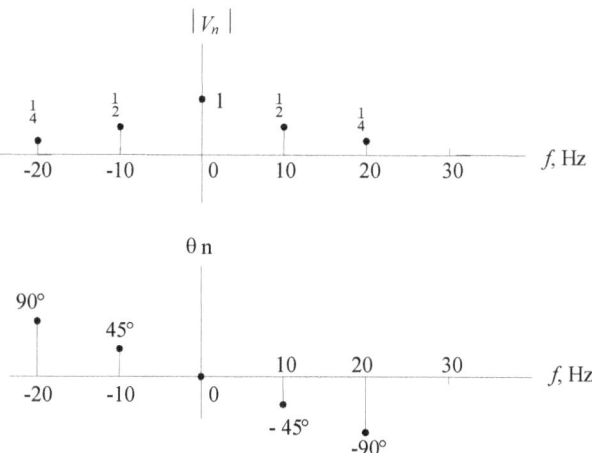

Fig. 7.5.1.

Problem 7.5.2. Repeat problem 7.5.1 for the spectrum shown in Fig. 7.5.2.

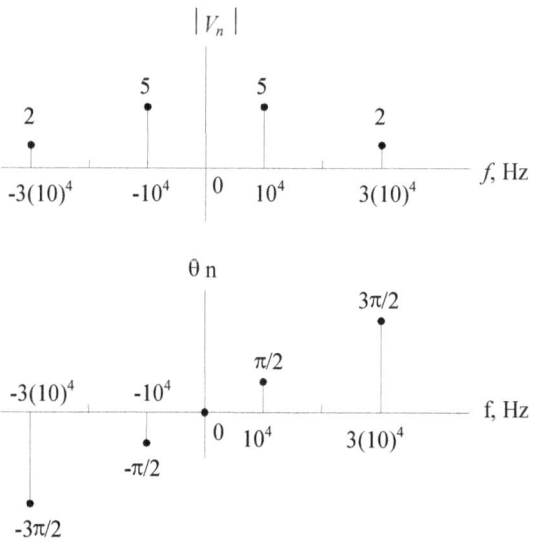

Fig. 7.5.2.

7.21

Answer:

$$v(t) = 10\cos\left(2\pi(10)^4 t + \pi/2\right) + 4\cos\left(6\pi(10)^4 t + 3\pi/2\right)$$

Self Test, objective 7.5.

Figure 7.5.3 shows the amplitude and phase spectrum of the periodic signal $v(t)$. Write an analytical expression for $v(t)$ in terms of trigonometric functions.

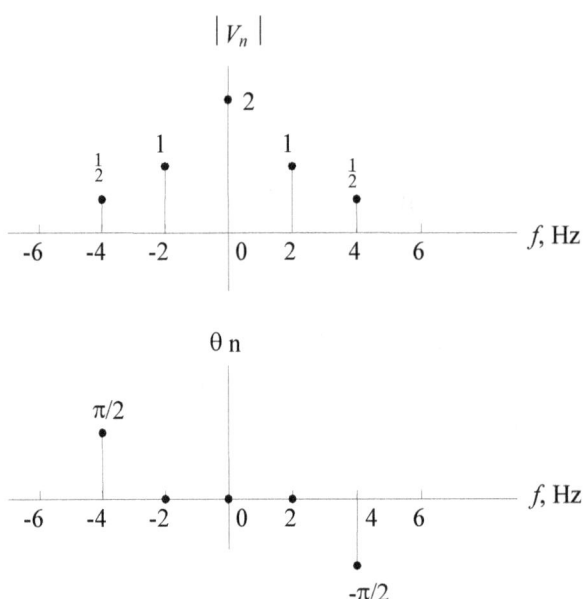

Fig.. 7.5.3.

Self Test Answers:

Objective 7.1.

Only number 3 is not a set of vectors. Matrices cannot be summed unless they are the same dimension, and therefore matrices with different dimensions cannot satisfy properties 1, 4, and 7 of the definition.

Objective 7.2.

The inner product is zero.

Objective 7.3.

1) See the strong Dirichlet conditions.

2) All have a Fourier series.

Objective 7.4.

1.

$$W_n = \frac{\Delta E}{T} \frac{\sin(n\pi\Delta/T)}{n\pi\Delta/T}$$

2.

$$V_n = \begin{cases} \dfrac{E}{2\pi n} e^{\frac{j\pi}{2}}, & n > 0 \\[2mm] \left|\dfrac{E}{2\pi n}\right| e^{\frac{j\pi}{2}}, & n < 0 \\[2mm] \dfrac{E}{2}, & n = 0 \end{cases}$$

Objective 7.5.

$$v(t) = 2 + 2\cos 4\pi + \cos(8\pi t - \pi/2)$$

Chapter 8

Response of LTI Systems by Fourier Series

Objectives: After completing this chapter you should be able to do the following:

8.1. Calculate the response of an LTI system to a signal expressed by its Fourier series.

8.2. Find and plot the power spectrum of a periodic waveform.

8.3. Calculate the average power on a one ohm basis in the output signal of an LTI system, where the input signal is periodic.

Rationale: This chapter combines the concepts of the previous two chapters. Chapter 6 determined the response of an LTI system to an eternal exponential signal. Chapter 7 decomposed a signal into its frequency components. If the signal was periodic, then the Fourier series represented the signal for all time, and the Fourier series is the sum of exponential signals. Thus objective 8.1 is simply a combination of these two ideas. It is one method for finding the response of LTI systems to periodic signals.

Objectives 8.2 and 8.3 introduce the concept of power in a signal. Many optimization problems are concerned with maximizing or minimizing power. A good example is a communication receiver where it is desirable to maximize the ratio of signal power to noise power. These objectives will give us a start in that direction.

Objective 8.1. Calculate the response of an LTI system to a signal expressed by its Fourier series.

Chapter 6 determined the response of an LTI system to an eternal exponential signal e^{st}. it is a simple matter to extend this concept to the sum of exponential signals by applying the LTI properties. The Fourier series is the sum of eternal exponential signals. We now combine these two concepts to find the response of LTI circuits.

Recall that the response to an exponential signal $e^{j\omega t}$ is $H(j\omega)e^{j\omega t}$. This is true for any frequency $\omega = \omega_1$. By the LTI properties, if the input signal is the sum of two exponential signals, say $e^{j\omega_1 t} + e^{j\omega_2 t}$, then the response is $H(j\omega_1)e^{j\omega_1 t} + H(j\omega_2)e^{j\omega_2 t}$. Hence, the response to a signal expressed by its Fourier series,

$$v(t) = \sum_{k=-\infty}^{\infty} V_k e^{j\omega_k t} \tag{8.1.1}$$

is given by

$$y(t) = \sum_{k=-\infty}^{\infty} H(jk\omega_1)V_k e^{j\omega_k t} \tag{8.1.2}$$

Therefore the procedure for calculating the response of LTI circuits is:

1. Calculate the Fourier coefficients for the input signal $v(t)$. That is, calculate V_k for each frequency $k\omega_1$.

2. Calculate the transfer function $H(j\omega)$.

3. Multiply $H(jk\omega_1)$ by V_k for each k.

4. Use Eq. 8.1.2 to compute the response $y(t)$.

Notes: (a) The magnitude of $H(jk\omega_1)$ is called the gain of the circuit at the frequency $\omega = k\omega_1$. This concept applies only to LTI circuits, and the gain is a function of frequency.

(b) The purpose of this section is to gain familiarity with the Fourier series. So don't look down your nose and say there are better ways to find the system output.

Ideal Filters: We will use the concept of ideal filters in the following problem. The filter characteristic for an ideal low-pass filter is plotted in Fig. 8.1.1a and that for an ideal band-pass filter in Fig. 8.1.1b. The ideal filter multiplies the input signal in the pass band by one, and all other frequency components are multiplied by zero.

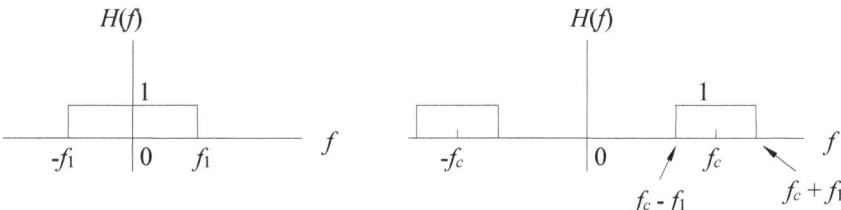

Fig. 8.1.1.

Notes: a) The bandwidth of the ideal low-pass filter is f_1 Hz.

b) The bandwidth of the band-pass filter is $2f_1$ and the center frequency is f_c.

c) The ideal filter is not physically realizable. (Why? See the discussion of the Paley-Wiener theorem in Chapter 12.)

Problem 8.1.1. The signal $v(t)$ shown in Fig. 8.1.2 is supplied to an ideal low-pass filter with cutoff frequency $f_1 = 18$ Hz. Find the output $y(t)$.

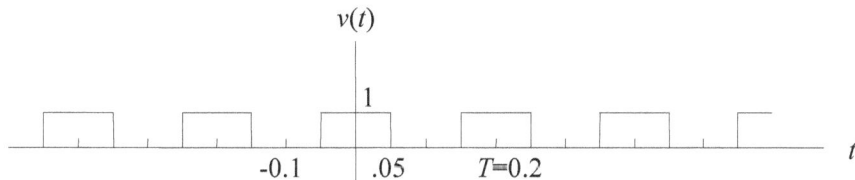

Fig. 8.1.2. The signal $v(t)$.

Solution: The first few terms of the series for $v(t)$ are plotted in Fig. 8.1.3, along with the ideal filter characteristic. From the diagram you can see that the output is given by

$$y(t) = -\frac{1}{3\pi}e^{-j3(2\pi)5t} + \frac{1}{\pi}e^{-j2\pi5t} + \frac{1}{2} + \frac{1}{\pi}e^{j2\pi5t} - \frac{1}{3\pi}e^{j3(2\pi)5t}$$

8.3

Or, in trigonometric form $y(t)$ is

$$y(t) = \frac{1}{2} + \frac{2}{\pi}\cos 2\pi 5t - \frac{2}{3\pi}\cos 2\pi 15t$$

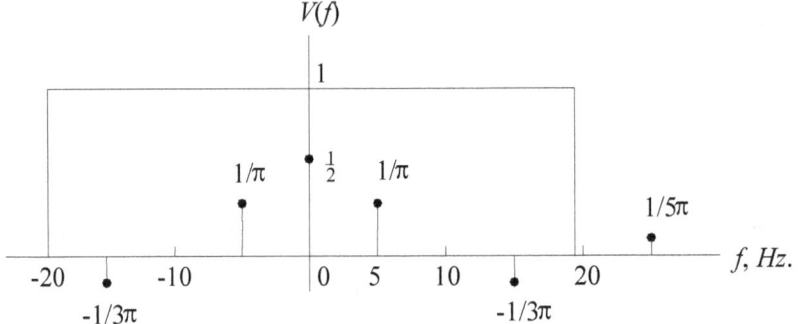

Fig. 8.1.3.

Problem 8.1.2. The signal $v(t)$ of the previous problem is now supplied to the RC low-pass filter shown in Fig. 8.1.4. Find the first few terms in the output $y(t)$.

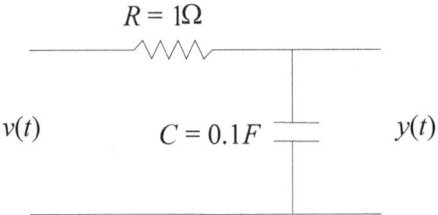

Fig. 8.1.4.

Solution: The filter characteristic is given by

8.4

$$H(f) = \frac{1}{1 + j2\pi f/10}$$

At frequencies f = 0, 5, 15 the value of $H(f)$ is shown in Table 8.1.1 along with the input components and the output components.

Table 8.1.1

f	$H(f)$	$V(f)$	$Y(f) = H(f)V(f)$
0	1	½	½
5	0.3e$^{-j72.4}$	$1/\pi$	0.0955e$^{-j72.4}$
-5	0.3e$^{j72.4}$	$1/\pi$	0.0955e$^{j72.4}$
15	0.106e^{-j84}	$-1/3\pi$	-0.01e^{-j84}
-15	0.106e^{j84}	$-1/3\pi$	-0.01e^{j84}

Thus the output y(t) is approximately given by

$$y(t) = \tfrac{1}{2} + 0.0955e^{-j2\pi 5t}e^{j72.4°} + 0.0955e^{j2\pi 5t}e^{-j72.4°}$$
$$- 0.01e^{-j3(2\pi)5t}e^{j84°} - 0.01e^{j3(2\pi)5t}e^{-j84°}$$

$$y(t) = \tfrac{1}{2} + 0.191\cos(10\pi t - 72.4°) - 0.02\cos(30\pi t - 84°)$$

Problem 8.1.3. The periodic square wave v(t) shown in Fig. 8.1.5 has period $T = 100\mu s$. This voltage is applied to an ideal low pass filter with 25 KHz bandwidth and gain of one. Express the output voltage y(t) as a function of time.

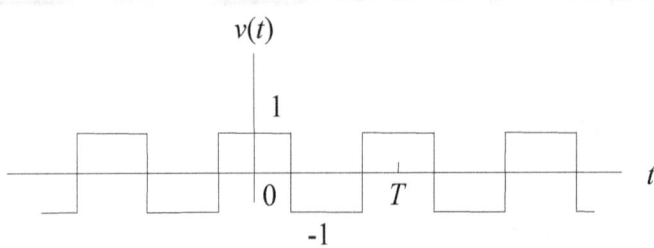

Fig. 8.1.5.

Answer: $y(t) = \frac{4}{\pi}\cos\left(2\pi(10)^4 t\right)$

Problem 8.1.4. The square wave in Problem 8.1.3 is now applied to the *RC* low pass filter shown in Fig. 8.1.6. Find the first few terms in the output $y(t)$.

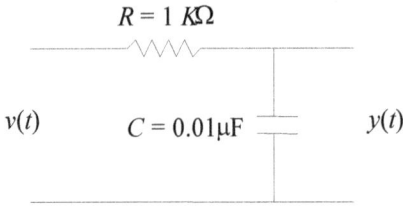

Fig. 8.1.6.

Answer:

$$y(t) \approx 1.08\cos\left[2\pi(10)^4 t - 32°\right] - 0.2\cos\left[6\pi(10)^4 t - 62°\right]$$
$$+ 0.076\cos\left[10\pi(10)^4 t - 72°\right]$$

Problem 8.1.5. The periodic square wave $v(t)$ shown in Fig. 8.1.7 has period $T = 100\mu s$. This voltage is applies to an ideal band pass filter with center frequency $f_c = 50KHz$ and bandwidth $B = 10KHz$. Find the output voltage $y(t)$.

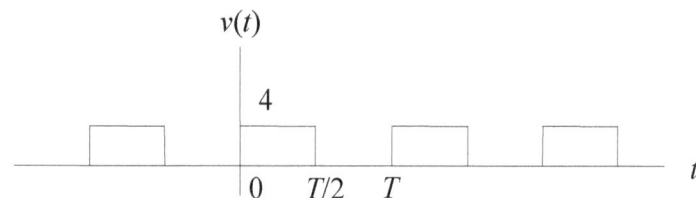

Fig. 8.1.7.

Solution: In each of the previous problems the input waveform was an even function of time. Hence the Fourier coefficients were real numbers. In Fig. 8.1.7 the function $v(t)$ is not even, and therefore the Fourier coefficients are complex numbers. The Fourier series is plotted in Fig. 8.1.8 as amplitude and phase. That is, we allow the amplitude to be either a positive or negative real number. Notice that this amplitude is the same (identical) to the Fourier series for an even function. Thus shifting the function $v(t)$ along the time axis serves to change only the phase spectrum. This is one of the properties of transforms that we will study in Chapter 12.

The fifth harmonic is the only output of the filter. Therefore the output is given by

$$y(t) = \frac{4}{5\pi}e^{-\frac{j5\pi}{2}}e^{j2\pi5(10)^4t} + \frac{4}{5\pi}e^{\frac{j5\pi}{2}}e^{-j2\pi5(10)^4t} = \frac{8}{5\pi}\cos(10^5\pi t - 5\pi/2)$$

8.7

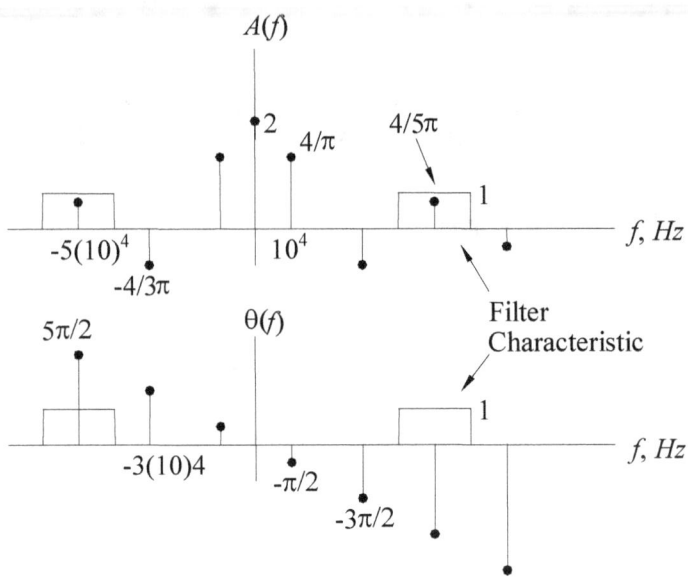

Fig. 8.1.8.

Problem 8.1.6. The physically realizable filter shown in Fig. 8.1.9 is now used in place of the ideal band-pass filter in Problem 8.1.5. Find the output $y(t)$.

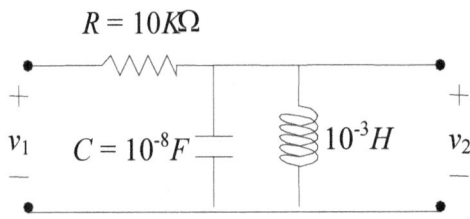

Fig.. 8.1.9.

Solution: We assume the inductor is ideal, although in practice there is always some loss associated with the inductor which must be accounted for by a resistor in the equivalent circuit. The impedance of the LC tank circuit is $Z(s)$ given by

$$Z(s) = \frac{sL/sC}{sL + 1/sC} = \frac{10^8 s}{s^2 + 10^{11}}$$

The transfer function is therefore given by

$$H(s) = \frac{Z(s)}{R + Z(s)} = \frac{10^4 s}{s^2 + 10^4 s + 10^{11}}$$

Set $s = j\omega$ to obtain

$$H(j\omega) = \frac{10^4 j\omega}{-\omega^2 + 10^4 j\omega + 10^{11}}$$

At the frequencies of interest the transfer function has the following values:

$$H\left(j2\pi 3(10)^4\right) = 0.0292e^{j90°}$$

$$H\left(j2\pi 5(10)^4\right) = 0.903e^{j25.6°}$$

$$H\left(j2\pi 7(10)^4\right) = 0.0354e^{-j90°}$$

Therefore, to a good approximation, the filter passes the input component at f_c with a gain of 0.903 and rejects the other harmonics.

Self Test, Objective 8.1. The periodic square wave shown in Fig. 8.1.10 is supplied to the RLC circuit. Find the approximate current in the resistor.

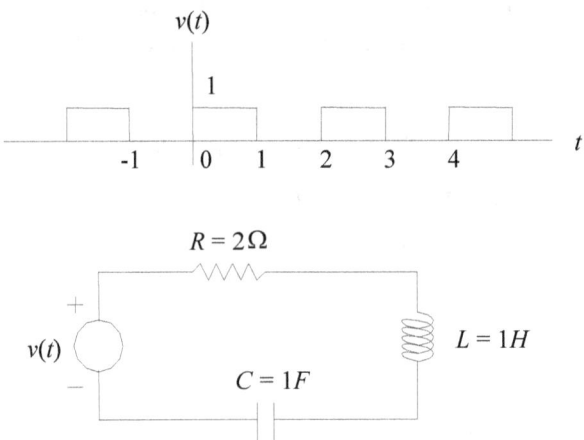

Fig. 8.1.10.

Objective 8.2. Find and plot the power spectral density function of a periodic waveform.

Classification of Signals

Signals may be classified into one of two broad categories:

1. Power signals.

2. Energy signals.

All practical signals (those that can be generated in the laboratory) can be classified into one of these two categories. Power signals have nonzero but finite power. Energy signals have nonzero but finite energy. Now to formalize these concepts:

Definition 8.2.1. Power Signals. The power in a signal $f(t)$ is given by

$$P = \lim_{a \to \infty} \frac{1}{2a} \int_{-a}^{a} |f(t)|^2 \, dt \qquad (8.2.1)$$

Any signal for which $0 < P < \infty$ is called a power signal.

Notes: a) If $P = 0$ the signal is not a power signal.

b) Equation 8.2.1 defines the average power in a signal on a one-ohm basis. Suppose the signal is current $i(t)$ that is flowing through a resistor R. Then the instantaneous power is given by $p(t) = i^2(t)R$, and the average power is

$$P = \lim_{a \to \infty} \frac{1}{2a} \int_{-a}^{a} |i(t)|^2 R \, dt$$

If $R = 1$ ohm this is Eq. 8.2.1.

c) If $f(t)$ is periodic with period T then Eq. 8.2.1 reduces to

$$P = \frac{1}{T} \int_{0}^{T} |f(t)|^2 \, dt$$

d) If $f(t)$ is a real function of time, then $|f(t)|^2 = f^2(t)$. If $f(t)$ is complex, then $|f(t)|^2 = f(t) \cdot f*(t)$ where $f*(t)$ denotes the complex conjugate of $f(t)$.

Definition 8.2.2. Energy Signals. The energy in a signal $f(t)$ is given by

$$E = \int_{-\infty}^{\infty} |f(t)|^2 \, dt \qquad (8.2.2)$$

Any signal for which $0 < E < \infty$ is called an energy signal.

Notes: a) Equation 8.2.2 is the total energy on a one ohm basis, just as Eq. 8.2.1 specified the average power in $f(t)$ on a one ohm basis.

b) The instantaneous power is related to the instantaneous energy $w(t)$ by

$$p(t) = \frac{d}{dt}w(t)$$

or, equivalently by

$$w(t) = \int_{-\infty}^{t} p(\lambda)d\lambda$$

where $p(t)$ and $w(t)$ are computed for the signal $f(t)$. There is no such simple relationship between average power and total energy.

c) Another name for energy signals is pulse signals.

Parseval's Theorem

Parseval's theorem for periodic signals relates the power in the time domain to the power in the frequency domain. For periodic signals the average power is given by

$$P = \frac{1}{T}\int_{t_1}^{t_1+T} |v(t)|^2 \, dt = \frac{1}{T}\int_{t_1}^{t_1+T} v(t)\cdot v^*(t)\, dt \qquad (8.2.3)$$

To show how this is related to the Fourier coefficients, start with Eq. 8.1.1.

$$v(t) = \sum_{k=-\infty}^{\infty} V_k e^{jk\omega_1 t} \qquad (8.1.1)$$

Substitute this into Eq. 8.2.3.

$$P = \frac{1}{T}\int_{t_1}^{t_1+T} v^*(t) \sum_{k=-\infty}^{\infty} V_k e^{jk\omega_1 t} \, dt \qquad (8.2.4)$$

In proving anything in which there is a double integration, a double summation, or an integration and a summation, the first thing to try (if permissible) is to swap the two integrals (or whatever). Thus in Eq. 8.2.4 interchange the integration and summation to get

$$P = \sum_{k=-\infty}^{\infty} V_k \left[\frac{1}{T} \int_{t_1}^{t_1+T} v^*(t) e^{j\omega_k t} dt \right]$$

The term in brackets is the complex conjugate of V_k. This leads to Parseval's theorem, given by

$$P = \frac{1}{T} \int_{t_1}^{t_1+T} |v(t)|^2 dt = \sum_{k=-\infty}^{\infty} |V_k|^2 \qquad (8.2.5)$$

The power can be found if the time function $v(t)$ is known. Likewise, the power can be found from the frequency function $V_k = V(f_k)$. As a simple example, consider the sinusoid

$$v(t) = \sin \omega_1 t \qquad (8.2.6)$$

The average power is

$$P = \frac{1}{T} \int_0^T \sin^2 \omega_1 t \, dt = \frac{1}{2}$$

The Fourier coefficients are $V_{-1} = \frac{1}{2} e^{-j\pi/2}$ and $V_1 = \frac{1}{2} e^{j\pi/2}$. Figure 8.2.1 shows the magnitude spectrum. All other coefficients are zero. This gives

$$P = \sum_{k=-\infty}^{\infty} |V_k|^2 = \frac{1}{4} + \frac{1}{4} = \frac{1}{2}$$

This is the same answer obtained in the time domain, as it should be.

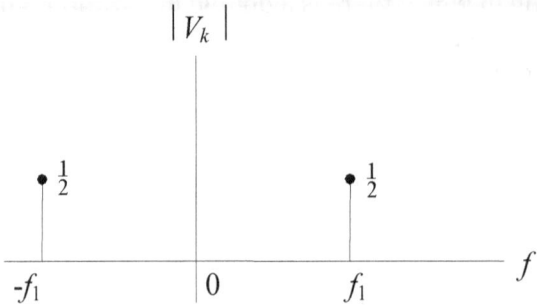

Fig. 8.2.1.

8.2.3. Power Spectral Density

Suppose an experiment is performed in the laboratory using the narrow band filter shown in Fig. 8.2.2. Only the magnitude of the transfer function is shown in the figure. The filter passes all frequency components in the narrow band of δf centered about f_0 and rejects all others. Suppose the filter is tunable, so that the center frequency f_0 is variable. (An approximation of such a filter is the RLC tank circuit that selects radio stations in an ordinary AM radio.)

A wattmeter is connected to the output of the filter. As the center frequency f_0 is varied throughout the frequency range the wattmeter reading changes. For the sinusoidal input in Eq. 8.2.6, for example, the wattmeter would read zero except when $2\pi f_0$ is close to ω_1. In the neighborhood of ω_1 the wattmeter will read 0.5 watts.

Suppose the input to the filter is a periodic square wave. Then the wattmeter will read zero except when $2\pi f_0$ is close to ω_1, $2\omega_1$, $3\omega_1$, etc. At these frequencies the wattmeter will read $2V_1^2$, $2V_2^2$, $2V_3^2$, etc. by Parseval's theorem. The V_i^2 terms are doubled because the filter is tuned to $+\omega_0$ and $-\omega_0$, thus passing components at both frequencies. (Of course there is no such thing as negative frequency in the laboratory, but the mathematics is greatly simplified if the double-sided spectrum is used.)

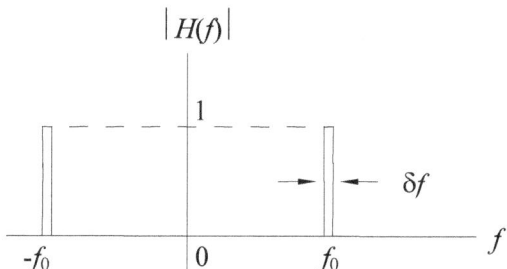

Fig. 8.2.2

As a third example, suppose the input to the filter is a general power signal, not necessarily periodic. The wattmeter reading will vary as the frequency f_0 is tuned from zero to infinity. A plot of wattmeter reading versus frequency such as Fig. 8.2.3 might result. Divide the wattmeter reading by two and represent one-half the power at positive frequencies and one-half at negative frequencies to make this plot. This curve contains information about the distribution of power along the frequency axis.

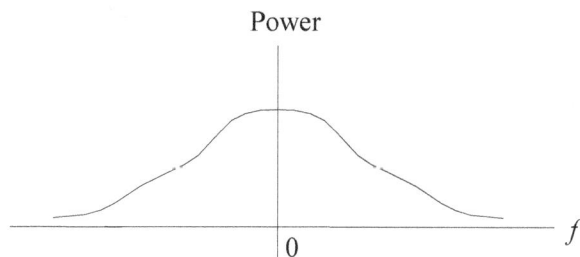

Fig. 8.23.

8.15

A curve such as that in Fig. 8.2.3 that specifies both the location and amount of power at each frequency is called the power spectral density function, $G(f)$. The total power in the frequency band $f_1 < f < f_2$ is given by

$$P_{f_1<f<f_2} = \int_{f_1}^{f_2} G(f)\,df \tag{8.2.7}$$

For the time being we are interested in the relationship between the Fourier coefficients V_k and the power spectral density. The power at any harmonic frequency f_k is just V_k^2. Since we wish to use integrals, we will represent the power at harmonic frequencies by delta functions. Therefore, for periodic signals the power spectral density $G(f)$ is given by a series of delta functions, and the areas under these delta functions are given by V_k^2.

An interesting and historically significant application of the concepts just outlined is contained in J. B. Johnson's noise experiments. Thermal energy causes agitation of electrons in a resistor. This thermal agitation, called Brownian motion, causes a small but measurable voltage to be impressed across the resistor, and it is this voltage that Johnson succeeded in measuring at Bell Labs in 1928.

This is called Brownian motion after the 19th century scientist Robert Brown who first observed the phenomena under a microscope while studying wheat pollen immersed in water. For an interesting account of the role played by Brown, Johnson, and many others including Albert Einstein, see D. K. C. MacDonald, "Noise and Fluctuations," John Wiley and Sons, 1962.

The principal behind Johnson's experiment is shown in Fig. 8.2.4. Since noise is present in all electronic circuitry, the input to the high gain amplifier is first shorted. The meter reading then indicates the rms voltage due to the high gain amplifier. When the resistor is connected to the amplifier input the additional meter reading is due to the thermal voltage across the resistor.

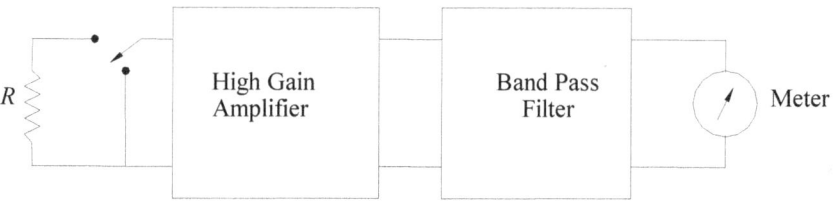

Fig. 8.2.4

Johnson found that the thermal voltage across a resistor was proportional to the square root of R and independent of frequency. That is, the thermal voltage changed when the bandwidth of the filter changed (δf in Fig. 8.2.2) but was independent of the center frequency f_0. Thus the power spectral density for this voltage is that shown in Fig. 8.2.5. The height is $2KTR$ where K is Boltzman's constant, T is the temperature of the resistor in degrees Kelvin, and R is the resistance in ohms.

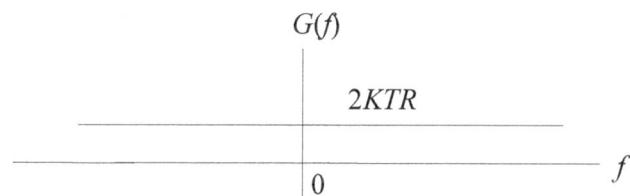

Fig. 8.2.5

Notes: a) This type of noise is termed white noise in analogy to white light. All frequencies are present.

b) Theoretically, white noise has a power spectral density that is constant for all frequency. Practically, quantum effects cause the curve to drop off at high frequencies.

c) It is this noise that limits communication. Electromagnetic signals grow weaker with distance, so they eventually become buried in the ever-present noise.

Problem 8.2.1. Find and plot the power spectral density function for the periodic signal in Fig. 8.2.26.

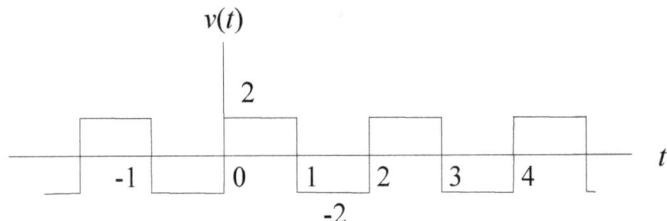

Fig. 8.2.6

Solution: See Fig. 8.1.8 for the Fourier series. Squaring each component gives the power spectrum in Fig. 8.2.7.

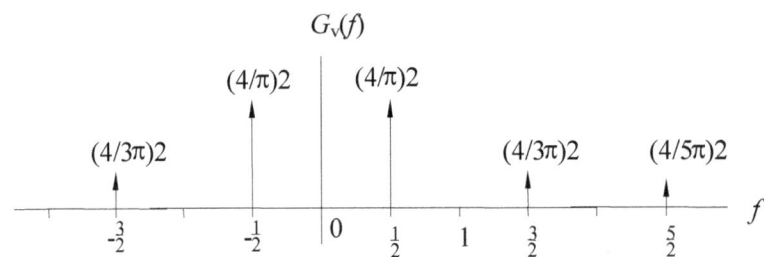

Fig. 8.2.7

Problem 8.2.2. Find and plot the power spectral density function for the triangular waveform in Fig. 8.2.8.

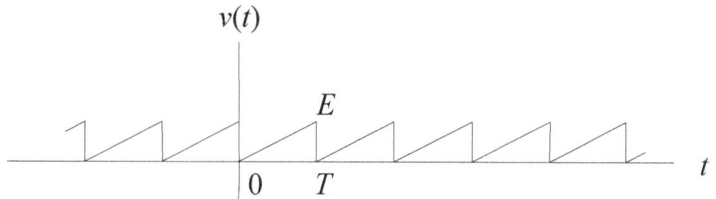

$v(t)$

E

$0 \quad T$

t

Fig. 8.2.8

Solution: The Fourier series for this waveform is given in the answer to the self test for objective 7.4. Squaring each component gives the power spectral density shown in Fig. 8.2.9.

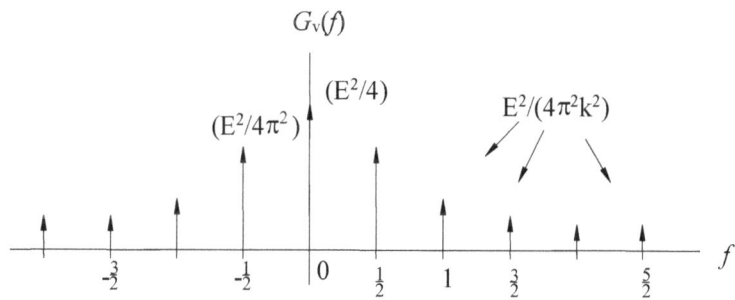

$G_v(f)$

$(E^2/4)$

$(E^2/4\pi^2)$

$E^2/(4\pi^2 k^2)$

$-\frac{3}{2} \quad -\frac{1}{2} \quad 0 \quad \frac{1}{2} \quad 1 \quad \frac{3}{2} \quad \frac{5}{2}$

f

Fig. 8.2.9

Self Test, Objective 8.2. Find and plot the power spectral density function for the waveform in Fig. 8.2.10.

8.19

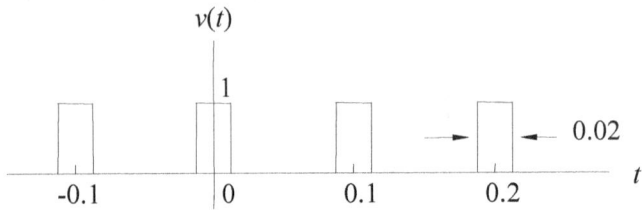

Fig. 8.2.10

Objective 8.3. Calculate the average power on a one ohm basis in the output signal of an LTI system, where the input signal is periodic.

Here the two previous objectives come together. Objective 8.1 showed how to find the frequency components in the output of an LTI system. From Objective 8.2 we know that these frequency components, when each is squared, form the power spectral density.

Problem 8.3.1. Calculate the average power on a one ohm basis for (a) the input signal and (b) the output signal in Problem 8.1.1.

Solution: (a) Use Eq. 8.1.1.

$$P = \frac{1}{0.2} \int_{-0.05}^{0.05} (1)^2 \, dt = 0.5 \; watts\,/\,ohm$$

(b) Square each component within the pass band in Fig. 8.1.3 and sum to obtain

$$P = \left(\tfrac{1}{2}\right)^2 + 2\left[\left(\tfrac{1}{\pi}\right)^2 + \left(\tfrac{1}{3\pi}\right)^2\right] \approx 0.475 \; watts\,/\,ohm$$

8.20

Problem 8.3.2. Calculate the average power in the output signal in Problem 8.1.2.

Solution: Square each component in Table 8.1.1 and sum to obtain

$$P = \left(\tfrac{1}{2}\right)^2 + 2\left[0.0955^2 + 0.01^2\right] \approx 0.432 \ watts / ohm$$

Self Test, Objective 8.3.

1. Find the total power (mean square value) of the waveform in Fig. 8.3.1.

2. If this voltage is supplied to the ideal low pass filter in the diagram, find the mean square value of the output.

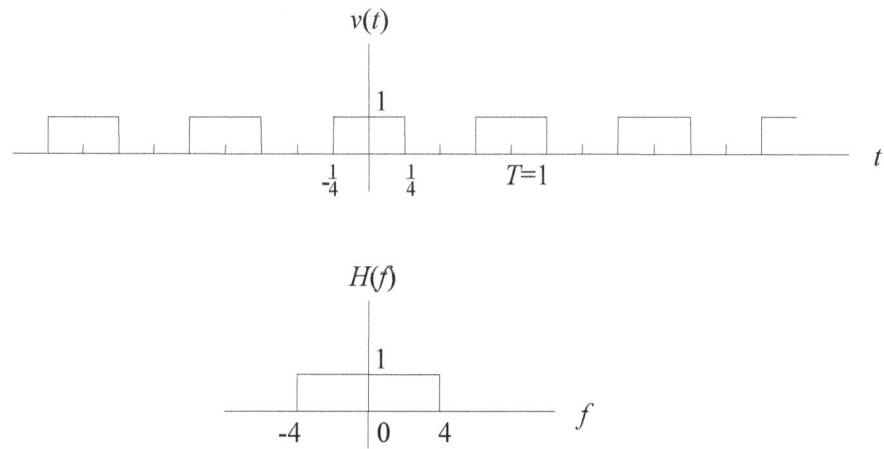

Fig. 8.3.1.

8.21

Self Test Answers

<u>Objective 8.1.</u>

$$i(t) \approx 0.182 \cos(\pi t - 145°) - 0.0224 \cos(3\pi t - 348°)$$

<u>Objective 8.2.</u>

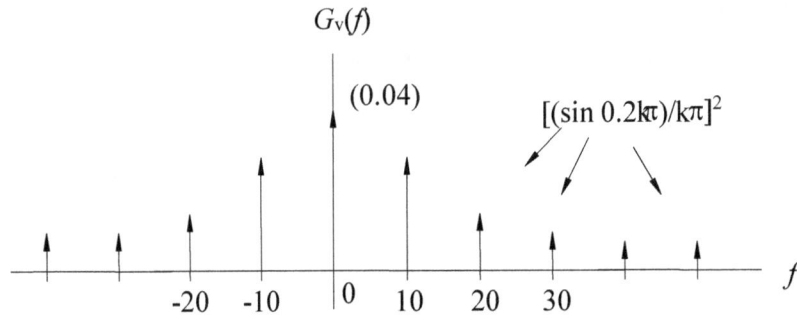

<u>Objective 8.3.</u>   1) 0.5,   2) 0.475

Chapter 9

Fourier Transform

Objectives: After completing this chapter you should be able to do the following:

9.1. Determine (select) which functions have a Fourier transform.
9.2. Find and plot the Fourier transform for a given time function.
9.3. Find and plot the time function corresponding to a given frequency function.

Rationale. Why should you learn about the Fourier transform?

For the same reasons you learned about the Fourier series. The series is a convenient way to express a periodic signal as the sum of exponential signals. In a similar way the transform is a convenient way to express a pulse-like signal as the sum of exponential signals. Then the LTI properties allow us to find the system response to these signals.

This chapter presents the mechanics of decomposing a pulse signal into its frequency components.

Objective 9.1. Determine (select) which functions have a Fourier transform.

9.1.1. The Fourier transform as an operator

The Fourier transform is an operator given by

$$V(f) = \int_{-\infty}^{\infty} v(t)e^{-j2\pi ft} dt \qquad (9.1.1)$$

Again, as with the Fourier series, this operator is pictured in Fig. 9.1.1. An element in the domain (a particular time function) is selected and fed into the function machine "Fourier transform." What comes out is the corresponding frequency function $V(f)$.

Fig. 9.1.1

Notes: a) $v(t)$ and $V(f)$ are themselves functions. In this application $v(t)$ is a function of time and $V(f)$ is a function of frequency. That is, the domain of v is a set of numbers that represent values of time, and the domain of V is a set of numbers that represent values of frequency.

b) The parameter f is continuous, in contrast to the parameter f_k for the Fourier series, which is discrete.

c) The range of both v and V can be a set of complex numbers. Since we deal with real signals the functions v will, for the most part, be real-valued.

The inverse operator is given by

$$v(t) = \int_{-\infty}^{\infty} V(f)e^{j2\pi ft} df \qquad (9.1.2)$$

Or, equivalently,

$$v(t) = \frac{1}{2\pi}\int_{-\infty}^{\infty} V(\omega)e^{j\omega t} d\omega \qquad (9.1.3)$$

Equation 9.1.3 is related to Eq. 9.1.2 by a simple change of variable, $\omega = 2\pi f$.

Note: As a general rule of thumb, whenever the variable of integration is ω, divide the integral by 2π. Whenever the variable of integration is f, do not divide by 2π. As with all rules of thumb, this must be used with caution, but at least one should guard against leaving off the 2π when integrating with respect to ω. This applies to all integrals, not just to the Fourier transform. For example Eq. 8.2.7 gives the power between two frequencies as

$$P_{f_1 < f < f_2} = \int_{f_1}^{f_2} G(f) df \qquad (8.2.7)$$

By the change of variable $\omega = 2\pi f$,

$$P_{\omega_1 < \omega < \omega_2} = \frac{1}{2\pi} \int_{\omega_1}^{\omega_2} G(\omega)d\omega \qquad (9.1.4)$$

The inverse Fourier transform is pictured in Fig. 9.1.2. The input to the function machine is selected from a set of frequency functions, and the corresponding output is a member of a set of time functions.

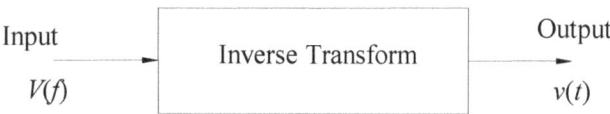

Fig. 9.1.2

9.1.2. Which Function Have a Fourier Transform?

The conditions that specify the domain of the forward operator are

$$\int_{-\infty}^{\infty} |v(t)| dt < \infty \qquad (9.1.5)$$

and

$$\int_{-\infty}^{\infty} |v(t)|^2 dt < \infty \qquad (9.1.6)$$

For $v(t)$ to have a Fourier transform given by Eq. 9.1.1 it is sufficient that Eqs. 9.1.5 and 9.1.6 be satisfied. If Eq. 9.1.6 is satisfied but Eq. 9.1.5 is not, then $v(t)$ has a transform, but it may not be given by Eq. 9.1.1.

Notes: a) Notice the similarity between the forward and inverse operators. The only difference is the sign on the exponent. Therefore those conditions that specify the domain of the inverse operator are given by Eqs. 9.1.5 and 9.1.6 with $V(f)$ substituted for $v(t)$ and integration over frequency instead of time.

b) If a function $v(t)$ satisfies Eqs. 9.1.5 and 9.1.6 then the Fourier transform $V(f)$ exists and

$$\int_{-\infty}^{\infty} |V(f)|^2\, df = \int_{-\infty}^{\infty} |v(t)|^2\, dt$$

c) Equation 9.1.6 is precisely our definition of energy signals. This is, of course, the reason for specifying the domain in this manner.

We now have methods to describe energy signals and periodic power signals in the frequency domain. As yet we have no such methods to describe nonperiodic power signals. This problem is encountered in random signal theory.

Self Test, Objective 9.1.

Determine which of the following functions have a Fourier transform.

1. $f(t) = \cos 2\pi t, \; -\infty < t < \infty$

2. $f(t) = \cos 2\pi t, \; 0 < t < 1$

3. The function shown in the first part of Fig. 9.1.3

4. The function shown in the second part of Fig. 9.1.3

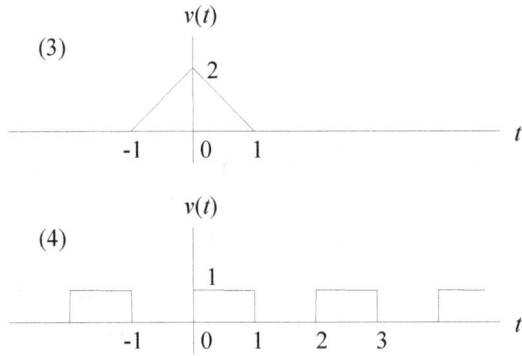

Fig. 9.1.3. Parts 3 and 4 of self test.

Objective 9.2. Find and plot the Fourier transform for a given time function.

For this objective we will simply apply the definition of Fourier transform, Eq. 9.1.1, to those functions that meet the criteria for having a Fourier transform. Here are some example problems.

Problem 9.2.1. Find the Fourier transform of $f_1(t)$ shown in Fig. 9.2.1. The function is given by

$$f_1(t) = \begin{cases} e^{at}, & t < 0 \\ e^{-at}, & t > 0 \end{cases}$$

Solution:

$$F_1(f) = \int_{-\infty}^{0} e^{at} e^{-j\omega t} dt + \int_{0}^{\infty} e^{-at} e^{-j\omega t} dt = \frac{1}{a - j\omega} + \frac{1}{a + j\omega} = \frac{2a}{a^2 + \omega^2}$$

Figure 9.2.1 shows the solution. Notice that $F_1(f)$ is a real function of frequency. This is true because $f_1(t)$ is even. But for most time functions the Fourier transform is a complex function of frequency. Therefore either a three-dimensional plot or two separate plots will graph this function.

Note: A function $f(t)$ is even if $f(t) = +f(-t)$. An odd function is one for which $f(t) = -f(-t)$.

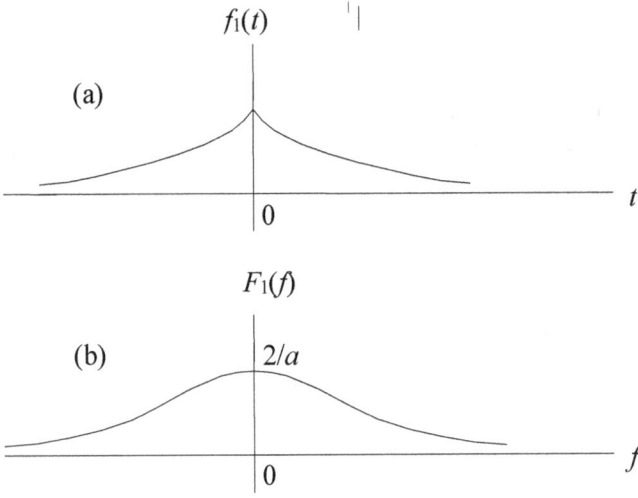

$f_1(t)$

(a)

0   t

$F_1(f)$

(b) $2/a$

0   f

Fig. 9.2.1

Problem 9.2.2. Find the Fourier transform of $f_2(t)$ shown in Fig. 9.2.2. The function $f_2(t)$ is given by

$$f_1(t) = \begin{cases} -e^{at}, & t < 0 \\ e^{-at}, & t > 0 \end{cases}$$

Solution:

$$F_1(f) = \int_{-\infty}^{0} -e^{at}e^{-j\omega t}\,dt + \int_{0}^{\infty} e^{-at}e^{-j\omega t}\,dt = \frac{-1}{a-j\omega} + \frac{1}{a+j\omega} = \frac{-j2a}{a^2+\omega^2}$$

Shown in Fig. 9.2.2, $F_2(f)$ is an imaginary function of frequency because $f(t)$ is odd.

Note: These two problems demonstrate one property of Fourier transforms. If $f(t)$ is even then $F(f)$ is real. If $f(t)$ is odd then $F(f)$ is imaginary.

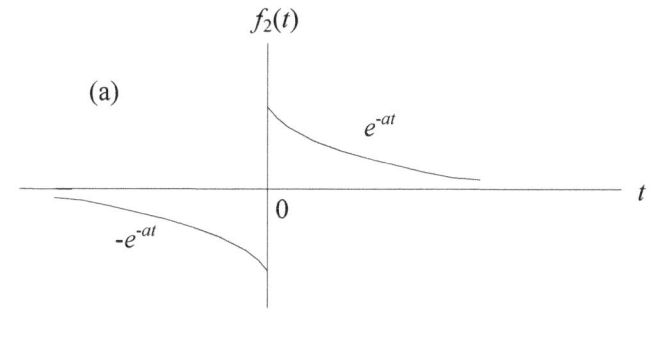

(a)

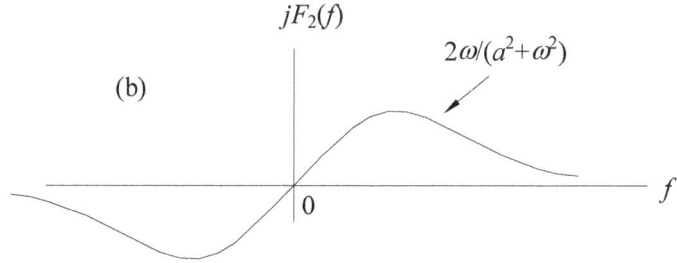

(b)

Fig. 9.2.2

Problem 9.2.3. Find the Fourier transform of the square pulse shown in Fig. 9.2.3.

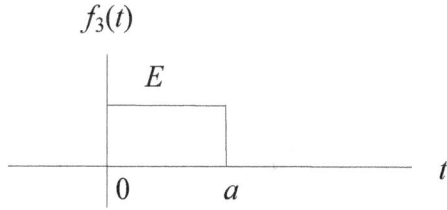

Fig. 9.2.3

Solution:

$$F_3(f) = \int_0^a Ee^{-j\omega t}dt = \frac{E}{j\omega}(1 - e^{-j\omega}) = Ea\left[\frac{\sin(\frac{a\omega}{2})}{a\omega/2}\right]e^{-\frac{ja\omega}{2}}$$

Since F_3 is complex, two plots are necessary as in Fig. 9.2.4. One plot graphs amplitude and the other graphs angle. F_3 is written in the following form:

$$F_3(f) = A_3(f)e^{j\theta_3(f)}$$

where $A_3(f)$ is the amplitude and $\theta_3(f)$ is the angle.

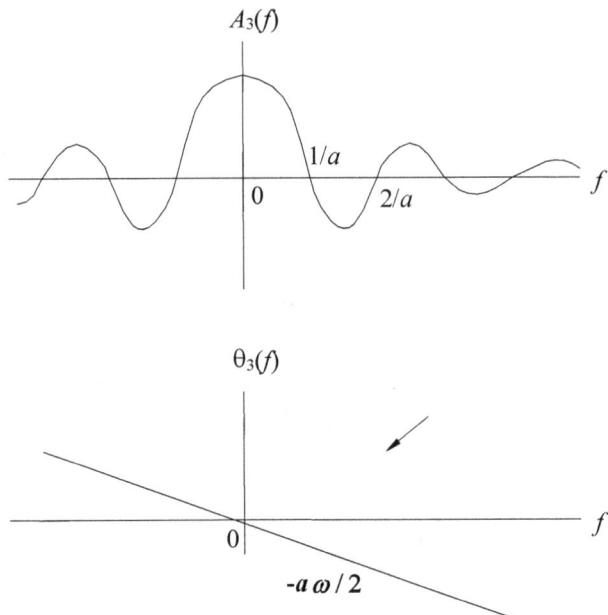

Fig. 9.2.4.

Problem 9.2.4. Find the Fourier transform of the periodic square wave in Fig. 9.2.5.

9.8

Solution: This waveform does not meet our criteria, so it has no Fourier transform in the usual sense. However, the Fourier transform does exist as distribution; that is, delta functions are present in the transform. We will discuss this problem after studying the properties of the transform where it will be convenient to combine our notation for the Fourier series and Fourier transform.

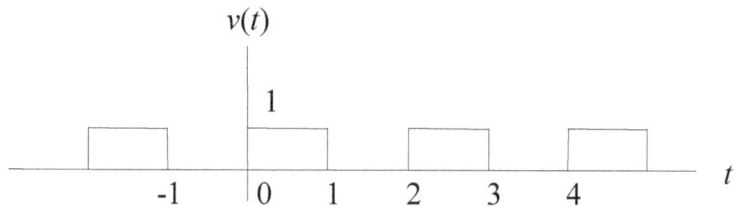

Fig. 9.2.5

Self Test, Objective 9.2.

Find the Fourier transform of the exponential function in Fig. 9.2.6.

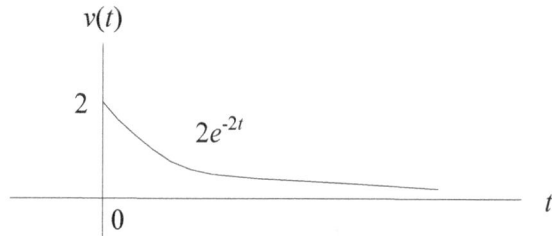

Fig. 9.2.6

Objective 9.3. Find and plot the time function corresponding to a given Fourier transform.

Here we apply Eqs. 9.1.2 or 9.1.3. Either of these are the inverse transform.

Problem 9.3.1. The Fourier transform of a function $f(t)$ is shown in Fig. 9.3.1. Find $f(t)$.

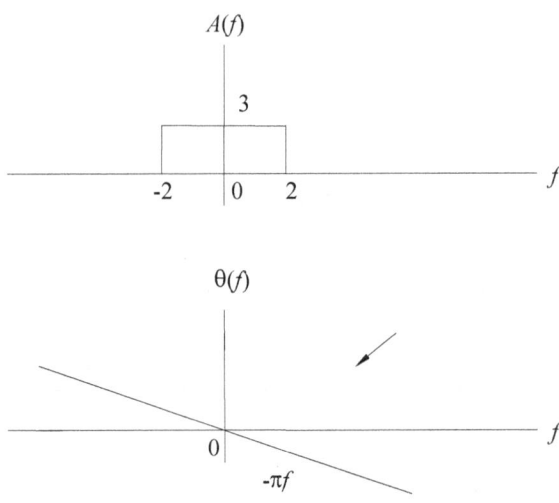

Fig. 9.3.1

Solution: The function $F(f)$ is given by

$$F(f) = A(f)e^{j\theta(f)} = \begin{cases} 0, & f < -2 \\ 3e^{-j\pi f}, & -2 < f < 2 \\ 0, & f > 2 \end{cases}$$

9.10

Apply Eq.. 9.1.2.

$$f(t) = \int_{-2}^{2} 3e^{-j\pi f} e^{j\omega t} df = \frac{3}{j(2\pi t - \pi)} \left[e^{j(2\pi t - \pi)2} - e^{-j(2\pi t - \pi)2} \right]$$

$$= 12 \left[\frac{\sin(4\pi t - 2\pi)}{4\pi t - 2\pi} \right]$$

Figure 9.3.2 shows a plot of this function.

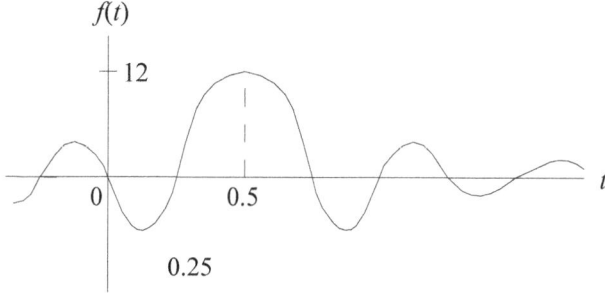

Fig. 9.3.2

Problem 9.3.2. Find the function $v(t)$ corresponding to the function $V(f)$ in Fig. 9.3.3.

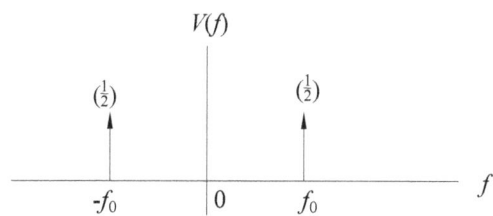

Fig. 9.3.3

9.11

Solution: Use Eq. 9.1.2.

$$v(t) = \int_{-\infty}^{\infty} \left[\frac{1}{2}\delta(f - f_0) + \frac{1}{2}\delta(f + f_0) \right] e^{j\omega t} df = \frac{1}{2}e^{j2\pi f_1 t} + \frac{1}{2}e^{-j2\pi f_1 t}$$

$$= \cos(2\pi f_0 t)$$

Note: This is an example of the special case discussed in Problem 9.2.4. The transform of a power signal $v(t)$ exists as distribution.

Problem 9.3.3. Find the function $y(t)$ corresponding to the function $Y(\omega)$ in Fig. 9.3.4.

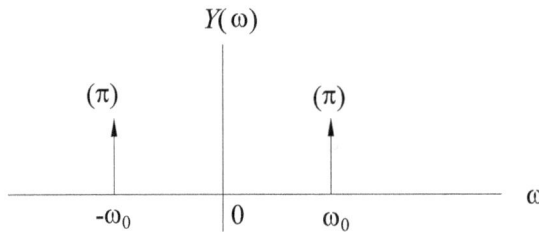

Fig. 9.3.4

Solution: Use Eq. 9.3.3.

$$v(t) = \frac{1}{2\pi} \int_{-\infty}^{\infty} [\pi\delta(\omega - \omega_0) + \pi\delta(\omega + \omega_0)] e^{j\omega t} dt = \cos(2\pi f_0 t)$$

Which is the same $v(t)$ in the previous problem.

Note: For a given function of frequency (f or ω) you must multiply the area under the δ-function by 2π when changing the variable from f in Hz to ω in radians per second.

Problem 9.3.4. Find $v(t)$ corresponding to $V(\omega)$ shown in Fig. 9.3.5.

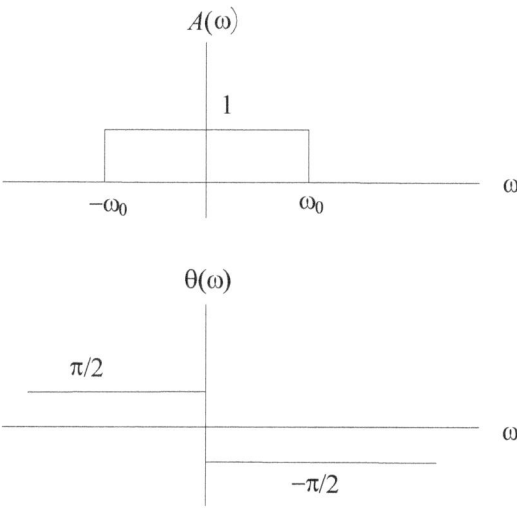

Fig. 9.3.5

Solution:

$$v(t) = \frac{1}{2\pi} \int_{-\infty}^{\infty} A(\omega) e^{j\theta(\omega)} e^{j\omega t} dt = \frac{1}{2\pi} \int_{-\omega_0}^{0} e^{\frac{j\pi}{2}} e^{j\omega t} dt + \frac{1}{2\pi} \int_{0}^{\omega_0} e^{-\frac{j\pi}{2}} e^{j\omega t} dt$$

$$= \frac{1}{\pi t} \left[\sin\left(\frac{\pi}{2}\right) + \sin(\omega_0 t - \frac{\pi}{2}) \right]$$

9.13

Self Test, Objective 9.3.

Find $v(t)$ corresponding to $V(f)$ shown in Fig. 9.3.6.

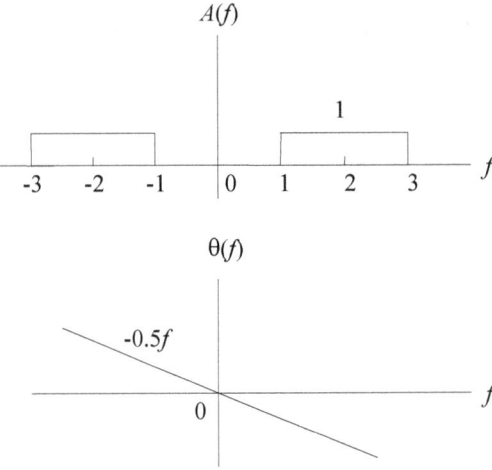

Fig. 9.3.6

Self Test Answers:

Objective 9.1. Functions 2 and 3 have a Fourier transform.

Objective 9.2. $V(\omega) = \frac{2}{2+j\omega}$

Objective 9.3.

$$v(t) = \frac{1}{j\left(2\pi t - \frac{1}{2}\right)}\left[e^{-j\left(2\pi t - \frac{1}{2}\right)} - e^{j\left(2\pi t - \frac{1}{2}\right)}\right]$$

$$+ \frac{1}{j\left(2\pi t - \frac{1}{2}\right)}\left[e^{j3\left(2\pi t - \frac{1}{2}\right)} - e^{-j3\left(2\pi t - \frac{1}{2}\right)}\right]$$

Appendix

A Table of Fourier Transform Pairs

You will need this table in the future. Also, we are not yet equipped to evaluate some of these – those for which $v(t)$ does not meet the requirements outlined in Section 9.1. We will learn to evaluate these pairs in Chapter 13.

$v(t)$	$V(\omega)$
$\delta(t)$	1
1	$2\pi\delta(\omega)$
$u(t)$	$\pi\delta(\omega)+\frac{1}{j\omega}$
$e^{-at}u(t)$	$\dfrac{1}{a+j\omega}$
$te^{-at}u(t)$	$\dfrac{1}{(a+j\omega)^2}$
$e^{-a|t|}$	$\dfrac{2a}{\left(a^2+\omega^2\right)}$
$\cos(\omega_0 t)$	$\pi\left[\delta(\omega+\omega_0)+\delta(\omega-\omega_0)\right]$
$\sin(\omega_0 t)$	$j\pi\left[\delta(\omega+\omega_0)-\delta(\omega-\omega_0)\right]$

Chapter 10

Response of LTI Systems by Fourier Transform

Objectives: After completing this chapter you should be able to do the following:

10.1. Calculate the response of an LTI system to a signal expressed by its Fourier Transform.

10.2. Find and plot the energy spectral density function for an energy signal.

10.3. Calculate the total energy on a one ohm basis in the output signal of an LTI system, where the input is an energy signal.

Rationale

We're doing the same thing in this chapter for the Fourier transform that we did in Chapter 8 for the Fourier series. The same rationale applies here. Just change the words "series" and "power" to "transform" and "energy" and re-read the rationale for LAP 8.

Objective 10.1. Calculate the response of an LTI system to a signal expressed by its Fourier Transform.

The procedures for calculating the response of LTI systems when the signal is expressed by the Fourier series and when the signal is expressed by the Fourier transform are analogous. For the Fourier series the sum is discrete. For the Fourier transform the sum is continuous and is given by

$$v(t) = \frac{1}{2\pi} \int_{-\infty}^{\infty} V(\omega) e^{j\omega t} dt \qquad (10.1.1)$$

The system response to a single exponential signal $V(\omega)e^{j\omega t}$ is given by $H(j\omega)V(\omega)e^{j\omega t}$. By the LTI properties the response to a signal expressed by the continuous sum in Eq. 10.1.1 is given by

$$v(t) = \frac{1}{2\pi}\int_{-\infty}^{\infty} H(j\omega)V(\omega)e^{j\omega t}d\omega \qquad (10.1..2)$$

You should compare these equations to Eqs. 8.1.1 and 8.1.2 in LAP 8. The following procedure is analogous to the one given there.

1. Calculate the Fourier transform for the input signal $v(t)$. That is, calculate $V(\omega)$.

2. Calculate the transfer function $H(j\omega)$.

3. Multiply the functions $V(\omega)$ and $H(j\omega)$.

4. Use Eq. 10.1.2 to compute the response $v(t)$.

Problem 10.1.1. An impulse $v(t) = \delta(t)$ is applied to the ideal low pass filter in Fig. 10.1.1. Find the response $y(t)$.

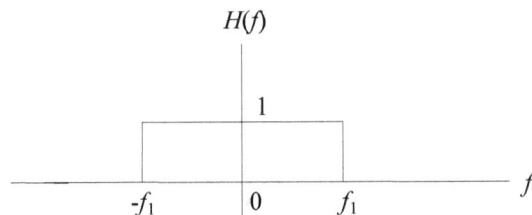

Fig. 10.1.1

Solution: First calculate $V(\omega)$.

$$V(\omega) = \int_{-\infty}^{\infty} \delta(t)e^{-j\omega t}dt = 1$$

Step 2 in the above procedure is provided by Fig. 10.1.1. For step 3 multiply $V(\omega)H(j\omega)$ given by.

$$V(\omega)H(\omega) = \begin{cases} 1, & |\omega| < \omega_1 \\ 0, & elsewhere \end{cases}$$

Then step 4 gives

$$y(t) = \frac{1}{2\pi} \int_{-\omega_1}^{\omega_1} 1e^{j\omega t}d\omega = \frac{\omega_1}{\pi}\left[\frac{\sin(\omega_1 t)}{\omega_1 t}\right] \quad -\infty < t < \infty$$

This response is plotted in Fig. 10.1.2. Notice that there is a response for $t < 0$. This says that a response occurs before the input is applied. Not good. This is the reason that the ideal filter is not physically realizable.

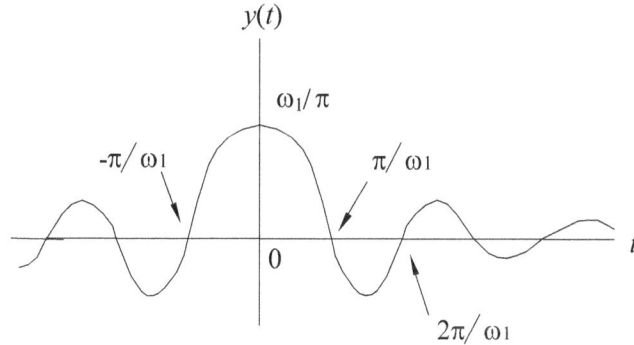

Fig. 10.1.2

Problem 10.1.2. Find the impulse response of the RC low pass filter shown in Fig. 10.1.3.

10.3

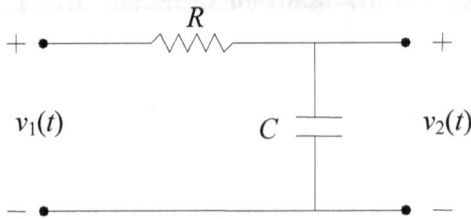

Fig.10.1.3

Solution: The transfer function $H(j\omega)$ is given by

$$H(j\omega) = \frac{1/RC}{j\omega + 1/RC}$$

Since the transform of $v(t) = \delta(t)$ is 1, the product $V(\omega)H(j\omega)$ is just $H(j\omega)$. Therefore the output is

$$v_2(t) = \frac{1}{2\pi} \int\limits_{-\infty}^{\infty} \frac{1/RC}{j\omega + 1/RC} e^{j\omega t} d\omega$$

We can avoid the difficulty in evaluating this integral with the table in LAP 9 Appendix. The relevant entry there is

$$e^{-at}u(t) \leftrightarrow \frac{1}{a + j\omega}$$

Therefore $v_2(t)$ is given by

$$v_2(t) = \frac{1}{RC} e^{-\frac{t}{RC}} u(t)$$

as shown in Fig. 10.1.4.

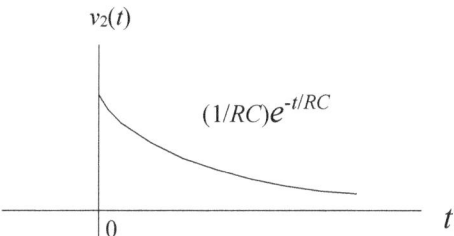

$v_2(t)$

$(1/RC)e^{-t/RC}$

0

t

Fig. 10.1.4

Note: This is just a simple introduction to the concept of finding the response by the Fourier transform. For any practical application we need the properties of transforms. That comes later.

Self Test, objective 10.1.

Find the current in the circuit if an impulse $v_1(t) = \delta(t)$ is applied to the circuit in Fig. 10.1.5.

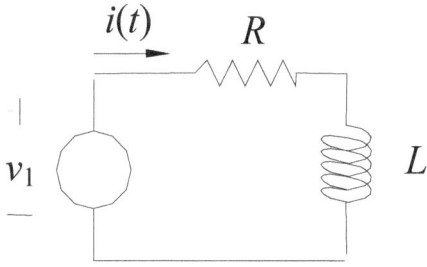

$i(t)$ R

v_1 L

Fig. 10.1.5

Objective 10.2. Find and plot the energy spectral density function for an energy signal.

10.2.1. Rayleigh's Theorem

Strictly speaking, the term "Parseval's theorem" refers to Eq. 8.2.3. Lord Rayleigh first used the corresponding relation for energy signals in his study of blackbody radiation. Rayleigh's theorem relates the energy in the frequency domain to the energy in the time domain, just as Parseval's theorem is a relation involving power. Recall that Eq. 8.2.2 gives the equation for energy as

$$E = \int_{-\infty}^{\infty} |v(t)|^2\, dt = \int_{-\infty}^{\infty} v^*(t)v(t)\, dt$$

Now replace $v(t)$ by its Fourier transform to obtain

$$E = \int_{-\infty}^{\infty} v^*(t) \left[\int_{-\infty}^{\infty} V(f)e^{j2\pi ft}\, df \right] dt$$

Again, as in Objective 8.2, change the order of integration to get

$$E = \int_{-\infty}^{\infty} V(f) \left[\int_{-\infty}^{\infty} v^*(t)e^{j2\pi ft}\, dt \right] df$$

The term in the brackets is $V^*(f)$, the complex conjugate of $V(f)$. Thus the energy is the same in the time domain as it is in the frequency domain.

$$E = \int_{-\infty}^{\infty} |v(t)|^2\, dt = \int_{-\infty}^{\infty} |V(f)|^2\, df \qquad (10.2.1)$$

This is known as Rayleigh's theorem.

10.2.2. Energy Spectral Density

Just as the power is distributed along the frequency axis for a power signal (see Objective 8.2), so must the energy be distributed along the frequency axis for an energy signal. That is, in dealing with an energy signal, we should be able to rig up some sort of apparatus to measure this energy in narrow frequency bands, and thus obtain a plot of energy versus frequency.

The energy spectral density function $W(f)$ is the function that (1) describes the relative amount of energy versus frequency, and (2) whose total area under $W(f)$ is the total energy. Thus we have

$$E = \int_{-\infty}^{\infty} W(f)\,df$$

The energy in the frequency band $f_1 < f < f_2$ is given by

$$E_{f_1 < f < f_2} = \int_{f_1}^{f_2} W(f)\,df$$

Now let's talk about the relationship between $W(f)$ and $V(f)$. The purpose of the following discussion is to show that

$$W(f) = |V(f)|^2 \qquad\qquad (10.2.2)$$

And the total energy in $v(t)$ is the area under $W(f)$. The energy density $W(f)$ is defined by two properties:

1. $W(f)$ describes the relative amount of energy versus frequency.

2. The total area under $W(f)$ is the energy.

The function $|V(f)|^2$ satisfies property 1, and Rayleigh's theorem states that the total area under $|V(f)|^2$ is the energy. We conclude that the function $|V(f)|^2$ fills the bill for the energy spectral density function. Hence Eq. 10.2.2 is valid.

Problem 10.2.1. Find and plot the energy spectral density function for the square pulse shown in Fig. 10.2.1a.

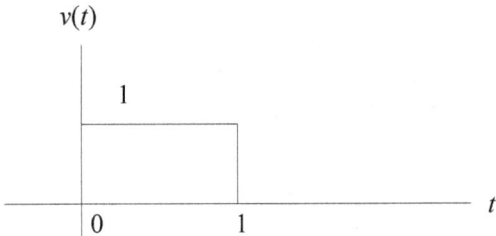

Fig. 10.2.1a

Solution: The Fourier transform of $v(t)$ is

$$V(f) = \int_{-\infty}^{\infty} v(t)e^{-j\omega t}\,dt = e^{-j\omega/2}\left[\frac{\sin(\omega/2)}{\omega/2}\right]$$

Square this function to obtain the distribution of energy.

$$|V(f)|^2 = \left[\frac{\sin(\omega/2)}{\omega/2}\right]^2$$

This is shown in Fig. 10.2.1b.

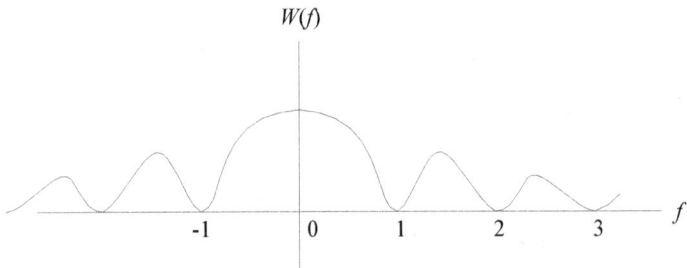

Fig. 10.2.1b

10.8

Self Test, Objective 10.2.

Find and plot the energy spectral density function W(f) for the exponential signal shown in Fig. 10.2.2.

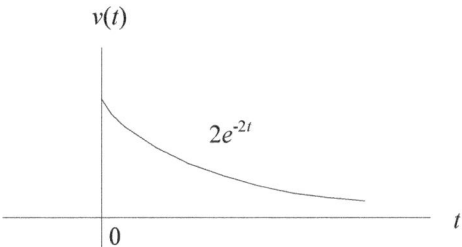

$v(t)$

$2e^{-2t}$

0

t

Fig. 10.2.2

Objective 10.3. Calculate the total energy on a one ohm basis in the output signal of an LTI system, where the input is an energy signal.

      The steps here are the same as for the Fourier series in Chap. 8. Here we combine the two previous objectives, as in Chap. 8, to find the energy in the output of an LTI system that has an energy signal input.

Problem 10.3.1. The square pulse $f(t)$ shown in Fig. 10.3.1 is supplied to an ideal low-pass filter with unit gain and bandwidth f_m. find the approximate energy in the output of the filter for the following values of f_m.

a) $f_m = 0.1\ Hz$
b) $f_m = 1\ Hz$
c) $f_m = 10\ Hz$

10.9

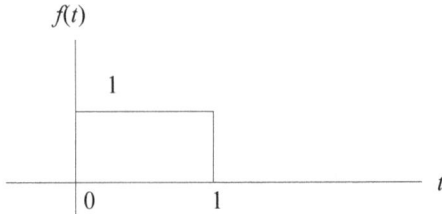

$f(t)$

1

0   1 t

Fig. 10.3.1

Solution: The energy spectral density function of the input signal is shown in Fig. 10.2.1b. The output signal is given by

$$Y(f) = H(f)F(f)$$

Therefore the output energy is given by

$$E_y = \int_{-\infty}^{\infty} |Y(f)|^2 df = \int_{-f_m}^{f_m} W_F(f) df$$

a) Here the limits on the above integral are (−0.1, 0.1). Since the magnitude of $W(f)$ at $f = 0$ is 1, the output energy is about 0.2 Joules/Ohm.

b) About 90% of the total energy is in the first lobe, from −1 < f < 1. Therefore the energy contained in the output signal is approximately 0.9 Joules/Ohm.

c) Almost all the signal energy is within this frequency range, so E_y is, for all practical purposes, 1 Joules/Ohm.

Self Test, Objective 10.3.

The exponential signal $v(t) = 2e^{-2t}u(t)$ shown in Fig. 10.2.2 is supplied to an ideal low-pass filter with gain 2 and bandwidth 0.1 *Hz*. Find the approximate energy in the output signal $y(t)$.

Self Test Answers:

<u>Objective 10.1.</u> $i(t) = \frac{1}{L} e^{-\frac{R}{L}t} u(t)$

<u>Objective 10.2.</u> $W(f) = \frac{4}{4+\omega^2}$ as shown in Fig. 10.3.2 below.

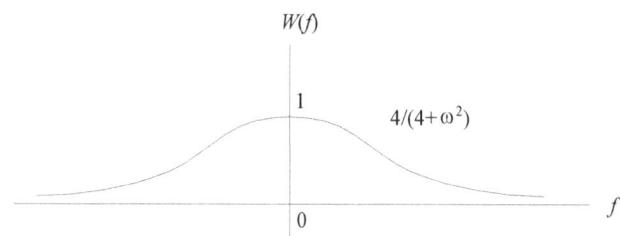

Fig. 10.3.2

<u>Objective 10.3.</u> E_y = 0.8 Joules/Ohm.

Chapter 11

Convolution

Objectives: After completing this chapter you should be able to do the following:

11.1. Find the impulse response of given LTI circuits.

11.2. Convolve two given functions.

11.3. Find the response of LTI circuits to given input signals by convolution.

Rationale: We have already introduced two methods of finding the response of LTI systems, differential equations and transforms. Convolution is the third of three methods. We interrupt our study of transform methods here because some of the properties of transforms depend on convolution. After this chapter we will return to the subject of transforms, and there we will need to have an understanding of convolution.

Chapter 11 Pre-Test

Use the following formula to evaluate the integral of each function in Pre-Test Diagram 1 below. Plot the result as a function of time.

$$y(t) = \int_{-\infty}^{t} x(\lambda)\,d\lambda$$

Check: You should be able to differentiate the function $y(t)$ in each case to obtain the function $x(t)$ given in the Pre-Test Diagram.

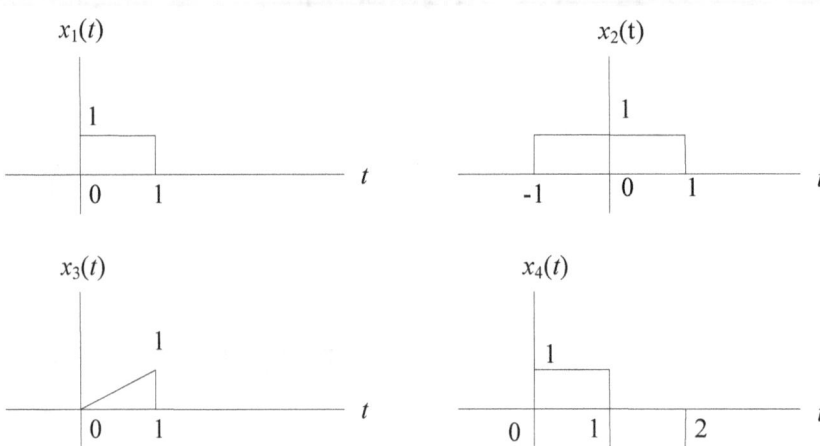

Pre-Test Diagram 1.

Objective 11.1. Find the impulse response of given LTI circuits.

The response of an LTI system to an impulse (the impulse response), labeled $h(t)$, is the transform of the transfer function $H(j\omega)$. This is written as

$$h(t) \leftrightarrow H(j\omega) \qquad (11.1.1)$$

To see this, consider an arbitrary LTI system with input $q(t)$ and output $y(t)$. If $q(t) = \delta(t)$ then the transform $Q(\omega)$ is a constant equal to 1. This means that $Y(\omega) = H(j\omega)Q(\omega) = H(j\omega)$. Therefore the impulse response is given by Eq. 11.1.1.

Problem 11.1.1. Find the impulse response of the RL low-pass filter in Fig. 11.1.1. The input is $v(t)$ and the output is $i(t)$.

Solution: The transfer function is given by

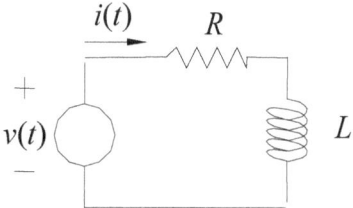

Fig. 11.1.1

$$H(s) = \frac{I(s)}{V(s)} = \frac{1}{R+sL} = \frac{1/L}{s+R/L}$$

Or
$$H(j\omega) = \frac{1/L}{j\omega+R/L}$$

The Table of Fourier Transform Pairs gives

$$h(t) = \frac{1}{L}e^{-\frac{R}{L}t}u(t)$$

Problem 11.1.2. Find the impulse response of the ideal low-pass filter whose transfer function is shown in Fig. 11.1.2.

Solution: First write an expression for $H(j\omega)$.

$$H(j\omega) = \begin{cases} 2e^{-j\omega}, & -\omega_C < \omega < \omega_C \\ 0, & elsewhere \end{cases}$$

Then calculate the inverse transform to find $h(t)$.

$$h(t) = \frac{1}{2\pi} \int_{-\omega_c}^{\omega_c} 2e^{-j\omega} e^{j\omega t} d\omega = \frac{2\omega_c}{\pi} \frac{\sin \omega_c(t-1)}{\omega_c(t-1)}$$

as shown in Fig. 11.1.3.

11.3

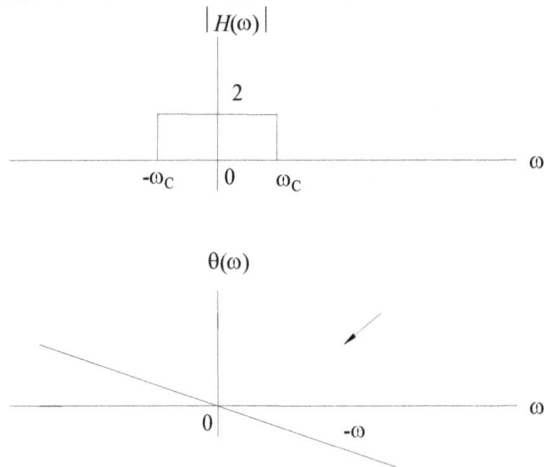

Fig. 11.1.2

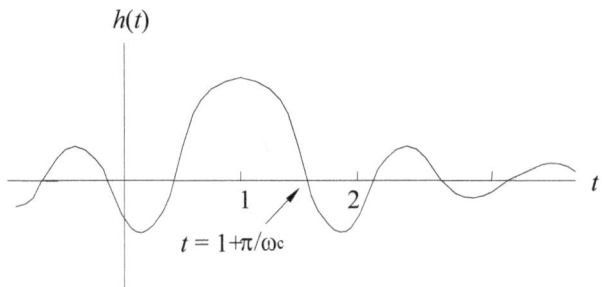

Fig. 11.1.3

Self Test, Objective 11.1.

Find the impulse response for each circuit shown in Fig. 11.1.4. The input is $v_1(t)$ and the output is $v_2(t)$ in each case.

11.4

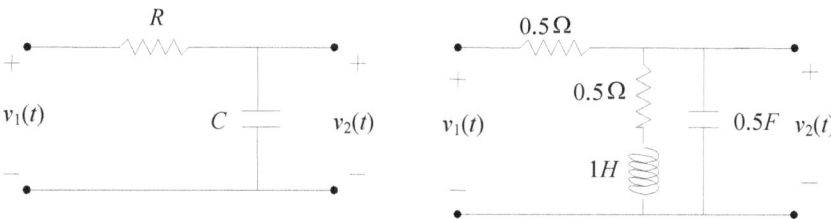

Fig. 11.1.4

Objective 11.2. Convolve two given functions.

Convolution is a binary operation. A binary operation is a mapping from $S{\times}S$ to S. Given a set S, select two elements from S (an ordered pair) and these two elements are used to produce a third element in the set S. For example, addition is a binary operation. If S is the set of real numbers then two numbers, say 2 and 4, are selected. Their sum produces a third element of S, the number 6.

A black box with two inputs and one output is a good illustration of this concept. The binary operation "addition" is shown in Fig. 11.2.1a where $c = a + b$, and "multiplication" is shown in Fig. 11.2.1b.

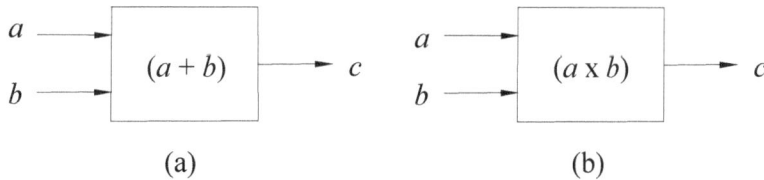

(a)   (b)

Fig. 11.2.1

There is no reason to restrict the inputs to be numbers. Convolution is a binary operation where the two inputs are functions and the output is a function. The symbol * is used for convolution just as + and × are used for addition and

11.5

multiplication. Figure 11.2.2 shows the black box that illustrates this binary operation.

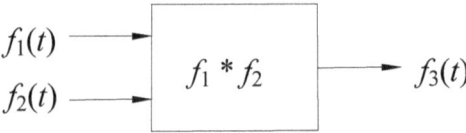

Fig. 11.2.2

The following equation defines convolution of two functions. Thus the two inputs to the black box are $f_1(t)$ and $f_2(t)$, and the output is $f_3(t)$.

$$f_3(t) = f_1(t) * f_2(t) = \int_{-\infty}^{\infty} f_1(\lambda)f_2(t-\lambda)d\lambda = \int_{-\infty}^{\infty} f_1(t-\lambda)f_2(\lambda)d\lambda \quad (11.2.1)$$

Let us begin with the simplest case and progress to more complex examples to show how to evaluate the integrals in Eq. 11.2.1.

Problem 11.2.1. Convolve the two step functions shown in Fig. 11.2.3.

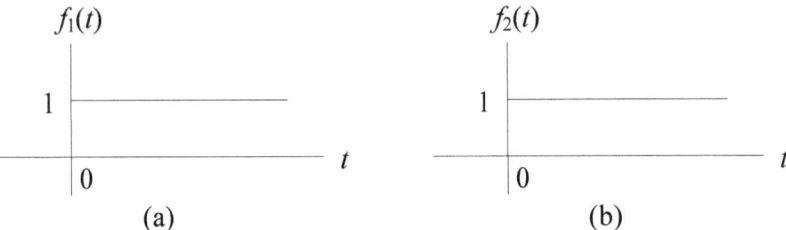

Fig. 11.2.3

Solution: The definition in Eq. 11.2.1 gives us a choice. Using the first version we must find $f_1(\lambda)$ and $f_2(t - \lambda)$. To find $f_1(\lambda)$ is easy, simply replace t by λ in Fig. 11.2.3.

Figure 11.2.4 shows the process of plotting $f_2(t - \lambda)$. It is usually best to plot a function of two variables versus the variable of integration, which in Eq. 11.2.1 is λ. Start by plotting $f_2(\lambda)$ versus λ in Fig. 11.2.4(a). Then "flip" this function to obtain $f_2(-\lambda)$ in (b).

Here is an important point. Notice that $f_2(-\lambda) = f_2(0 - \lambda)$, so this is $f_2(t - \lambda)$ if $t = 0$. This same function is shown for other values of t in (c) and (d), first for a negative value of t (say $t = -1$) and then for a positive value of t.

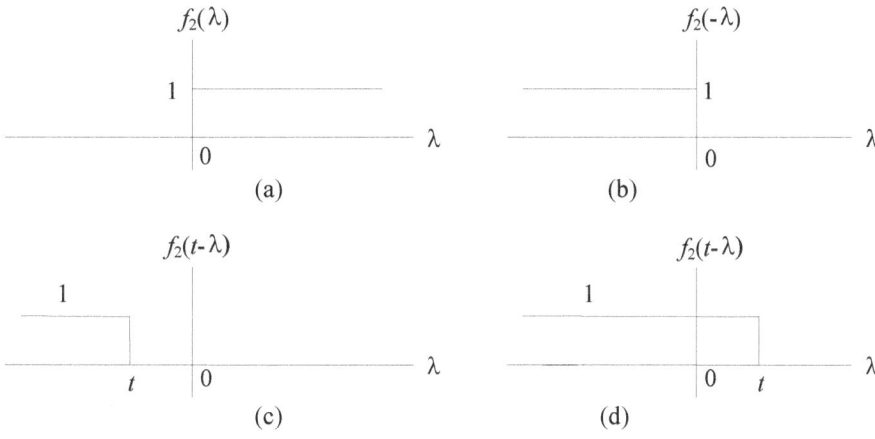

Fig. 11.2.4

Now multiply $f_1(\lambda)$ by $f_2(t - \lambda)$ and integrate according to Eq. 11.2.1. This must be done for every value of t in the interval $-\infty < t < \infty$. Figure 11.2.5 continues the solution. In (a) the value of t is less than zero and the product $f_1(\lambda)f_2(t - \lambda)$ is zero for every value of λ. In (b) with $t > 0$ the product $f_1(\lambda)f_2(t - \lambda)$ is equal to 1 for $0 < \lambda < t$ and 0 elsewhere. (Here you can see the value of plotting $f_2(t - \lambda)$ versus λ.) Thus $f_3(t)$ is given by

$$f_3(t) = \int_0^t 1\,dt = t, \quad t > 0$$

as shown in (c).

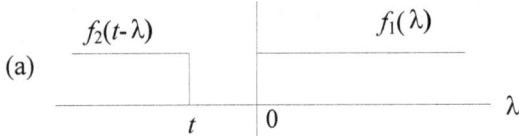

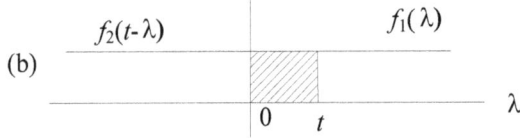

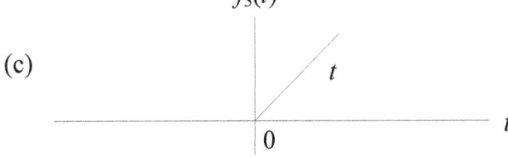

Fig. 11.2.5

Problem 11.2.2. Convolve the two functions in Fig. 11.2.6.

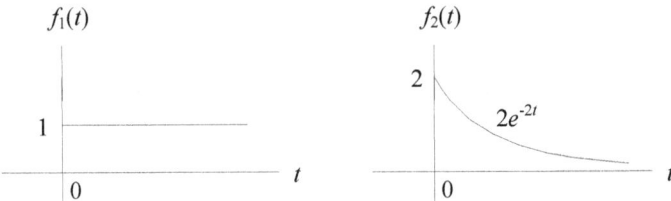

Fig. 11.2.6

11.8

Solution: Let's flip and slip $f_2(t)$ to show how this is done. (It would be easier to flip and slip $f_1(t)$.) Figure 11.2.7 below proceeds from $f_2(\lambda)$ in (a) to the evaluation of $f_3(t)$ for different values of t in (c) and (d). The integral in (d) is

$$f_3(t) = \int_0^t 2e^{-2(t-\lambda)}d\lambda = 1 - e^{-2t}, \ \ t > 0$$

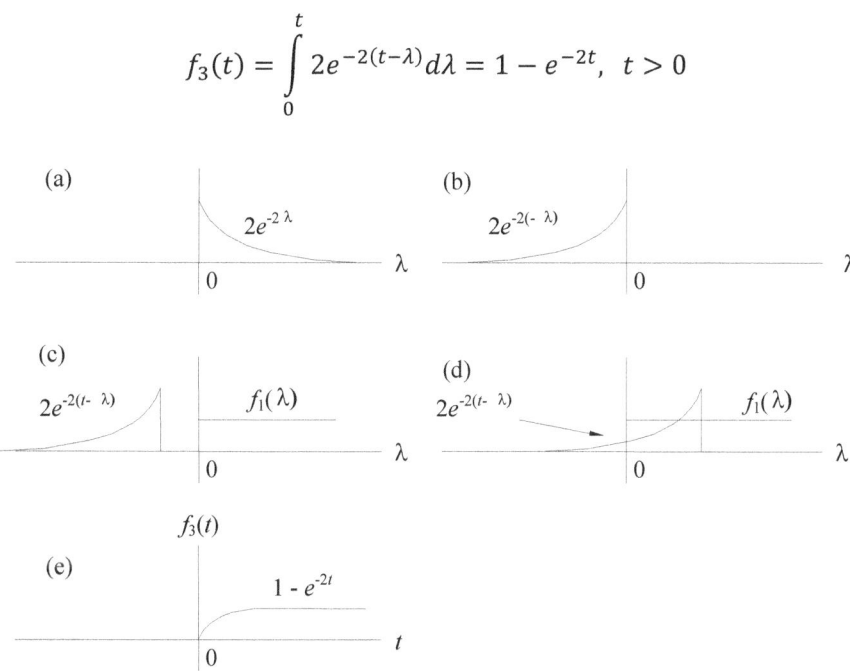

Fig. 11.2.7

One thing you must be able to do is write the equation for $f(t - \lambda)$. The next problem illustrates this with several different functions.

Problem 11.2.3. For each function $f(t)$ shown in Fig. 11.2.8, plot $f(t - \lambda)$ for a value of t given by $t = 1.5$. Also label the figures with the correct equation.

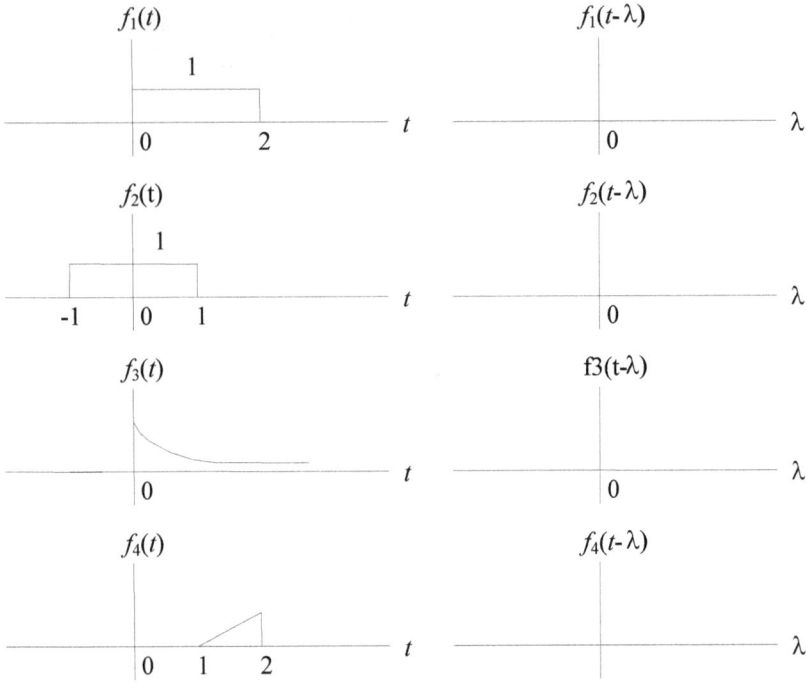

Fig. 11.2.8

Solution: Each graph in Fig. 11.2.9 shows both $f(-\lambda)$ and then $f(t-\lambda)$ for $t = 1.5$.

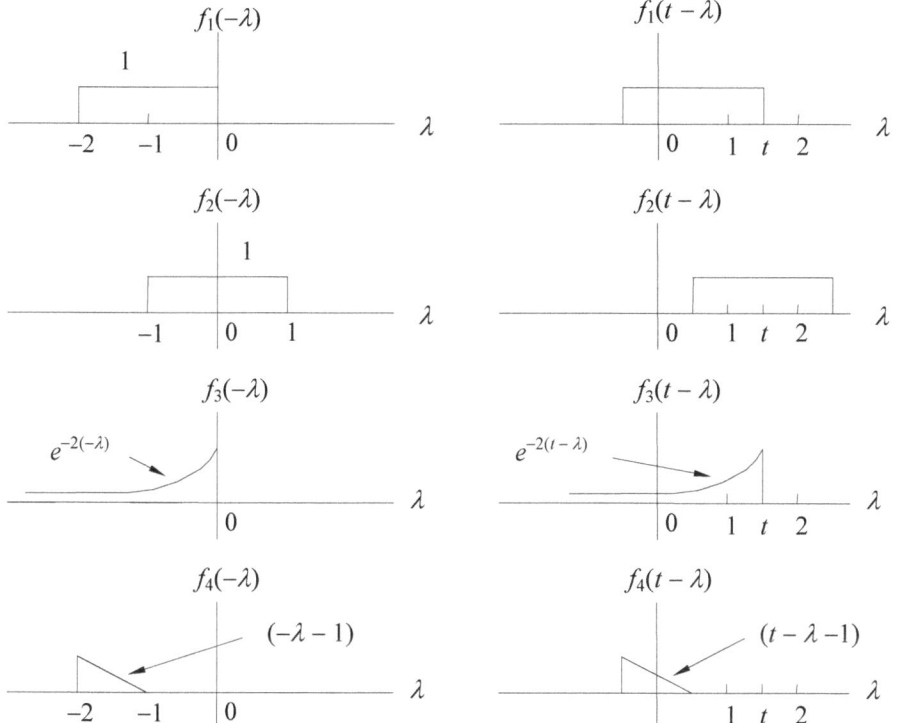

Fig. 11.2.9

Here is the procedure for evaluating convolution integrals:

1. Plot f_1 and f_2 as functions of λ rather than t.
2. Select one function to flip and slip, say f_2.
3. flip $f_2(\lambda)$ to obtain $f_2(-\lambda)$.
4. Slip f_2 to left or right until the point originally at the origin coincides with the present value of t.
5. The area under the product $f_1(\lambda)f_2(t - \lambda)$ is the value of $f_3(t)$ for that one value of t.
6. vary t from $-\infty$ to ∞.

Self Test, Objective 11.2.

Convolve the two functions in Fig. 11.2.10.

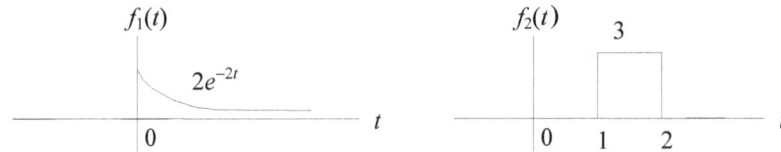

Fig. 11.2.10

Objective 11.3. Find the response of LTI circuits to given input signals by convolution.

Our purpose here, before getting to our objective, is to derive the convolution integral by direct application of the LTI properties. This illustrates that the system output $y(t)$ is related to the system input $q(t)$ by convolution with the impulse response $h(t)$. That is, $y(t)$ is given by

$$y(t) = \int_{-\infty}^{\infty} q(\lambda)h(t-\lambda)\,d\lambda \qquad (11.3.1)$$

The system is pictured in Fig. 11.3.1.

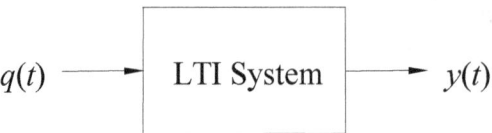

Fig. 11.3.1

To begin, assume that the input to an arbitrary LTI system is a unit step $u(t)$. Then the output is the step response $s(t)$. Figure 11.3.2(a) shows the input $q(t)$ along with a general depiction of the output $y(t)$. Notice from the diagram that the output at time t_a is $y(t_a) = s(t_a)$ The second part of the diagram shows the input delayed by time t_j. Because of LTI the output is also delayed. Now the response at time t_a is given by $y(t_a) = s(t_a - t_j)$.

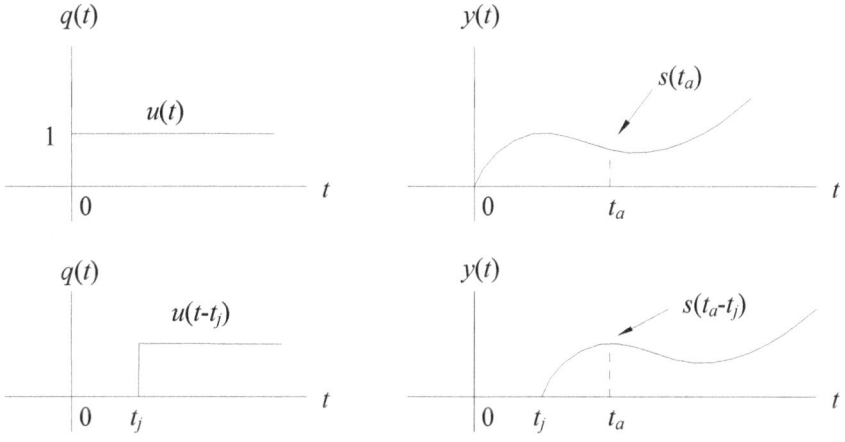

Fig. 11.3.2(a)

Having established that the response at time t_a to a unit step delayed by t_j is $s(t_a - t_j)$, consider the arbitrary input signal $q(t)$ in Fig. 11.3.2b. This signal can be approximated by a series of square pulses, where each pulse is the sum of two step functions. The square pulse at time t_j is given by $u(t - t_j) - u(t - t_j - \Delta t)$. Then the input $q(t)$ can be approximated by

$$q(t) = \sum_{j=0}^{N} q(t_j)[u(t - t_j) - u(t - t_j - \Delta t)]$$

11.13

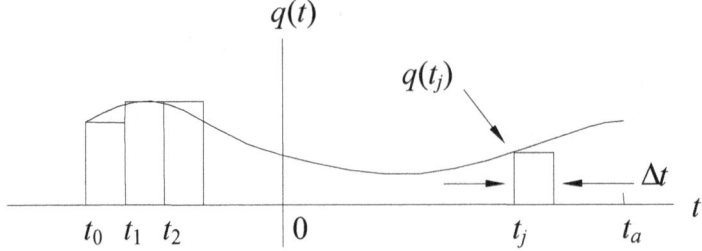

Fig. 11.3.2(b)

The response at time t_a to a single square pulse, say the one shown in the diagram at t_j, is given by

$$y(t_a) = q(t_j)[s(t_a - t_j) - s(t_a - t_j - \Delta t)]$$

Apply the superposition property to approximate the response to $q(t)$ at time t_a.

$$y(t_a) = \sum_{j=0}^{N} q(t_j)\left[s(t_a - t_j) - s(t_a - t_j - \Delta t)\right]$$

Furthermore, for small Δt this is approximately

$$y(t_a) = \sum_{j=0}^{N} q(t_j) h(t_a - t_j) \Delta t$$

Where did this come from? First, $h(t)$ is the derivative of $s(t)$. Thus the impulse response is given approximately by

$$h(t_a - t_j) \cong \frac{s(t_a - t_j) - s(t_a - t_j - \Delta t)}{\Delta t}$$

Now if we let $\Delta t \to 0$ in this expression the approximation becomes exact and $y(t_a)$ is given by

$$y(t_a) = \int_{t_0}^{t_a} q(\lambda)h(t_a - \lambda)d\lambda \qquad (11.3.2)$$

In general, the limits on the convolution integral are infinite. They are finite here because $q(t) = 0$ for $t < t_0$ and $h(t) = 0$ for $t < 0$, or $h(t_a - \lambda) = 0$ for $\lambda > t_a$.

Notes: a) We began this discussion by characterizing the system by the $u(t)$, $s(t)$ pair. The convolution integral was derived from this pair by applying the linearity and time-invariance properties, thus showing that convolution applies to LTI systems.

b) Convolution applies to any LTI system impressed with any input signal $q(t)$ that can be represented by the sum of narrow pulses. Any $q(t)$ that can be generated physically can be so represented.

Self Test, Objective 11.3.

Use convolution to find the response of the circuit in Fig. 11.3.3 to the signal shown.

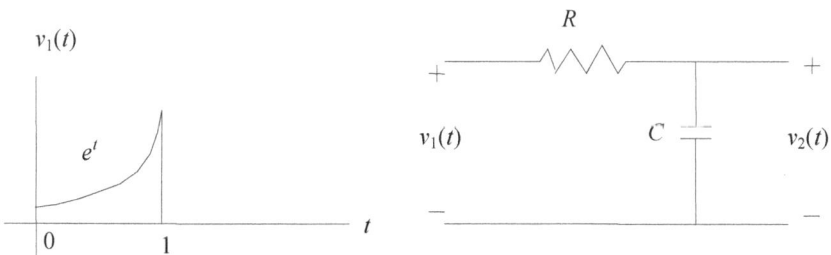

Fig. 11.3.3

Pre-Test Answer

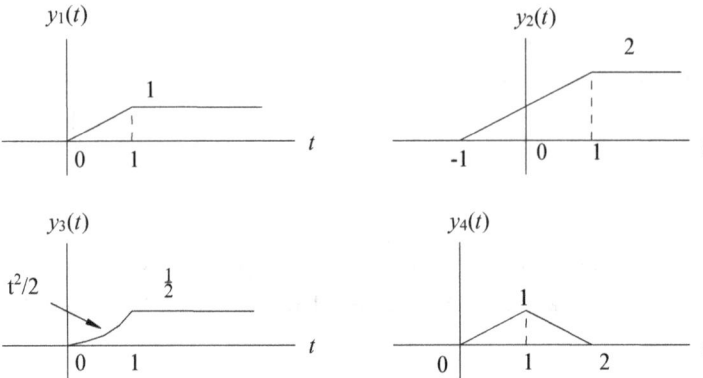

Pre-Test Diagram 2.

Self Test Answers:

<u>Objective 11.1.</u>

$$h_1(t) = \frac{1}{RC} e^{-t/RC} u(t)$$

$$h_2(t) = (8e^{-3t} - 4e^{-2t})u(t)$$

<u>Objective 11.2.</u>

$$f_3(t) = \begin{cases} 0, & t < 1 \\ 3[1 - e^{2(1-t)}], & 1 < t < 2 \\ 3[e^4 - e^2]e^{-2t}, & 2 < t \end{cases}$$

<u>Objective 11.3.</u>

$$v_2(t) = \begin{cases} 0, & t < 0 \\ \dfrac{1}{2}[e^t - e^{-t}], & 0 < t < 1 \\ \dfrac{e^{-t}}{2}[e^2 - 1], & 1 < t \end{cases}$$

Chapter 12

Properties of Fourier Transform

Objectives: After completing this chapter you should be able to do the following:

12.1. Use the differentiation, delay, modulation, convolution, and multiplication properties of Fourier transforms to evaluate transforms of given functions.

12.2. Find bounds on the spectrum of a given time function by using the concepts of content, variation, and wiggliness.

12.3.Use the Paley-Wiener theorem to test the magnitude of transfer functions for physical realizability.

Rationale: Why study the properties of transforms?

It is the properties of Fourier (and Laplace) transforms that make them useful. For example, it is absolutely essential that you have some knowledge of these properties in order to understand the operation of the following:

1. Communication systems (AM, FM, PCM, etc.)

2. Chopper stabilized dc amplifiers.

3. Holography and other concepts in optics.

4. Any signal-processing task as in the exploration for oil, detection of submarines, satellite tracking, etc. And there are many other applications. For example, the Heisenberg uncertainty principle was developed using a property of Fourier transforms. The French mathematician Fourier originally developed the transform in his study of heat.

Objective 12.2, concerned with bounds on the spectrum, is included in this chapter to illustrate that the location of a signal spectrum along the frequency axis can easily be approximated without actually finding the transform.

Objective 12.3 introduces the Paley-Wiener theorem, which is concerned with the physical realizability of transfer functions. Given the transfer function, the Paley-Wiener theorem determines if this function could possibly be physically realizable.

Objective 12.1. Use the differentiation, delay, modulation, convolution, and multiplication properties of Fourier transforms to evaluate transforms of given functions.

12.1.1. A list of properties

Some of the more useful properties of the Fourier transform are listed below. These properties for the transform are identical to those for the series, with appropriate modifications. (Essentially, this means substituting the discrete parameter ω_k or f_k for the continuous parameter ω or f in the transform.)

Uniqueness: $v(t) \leftrightarrow V(f)$

Superposition: $av(t) + bw(t) \leftrightarrow aV(f) + bW(f)$

Differentiation: $\frac{d^n v(t)}{dt^n} \leftrightarrow (j\omega)^n V(f)$

Scaling: $v(t/a) \leftrightarrow aV(af) \quad \text{for } a > 0$

Delay: $v(t - t_0) \leftrightarrow V(f)e^{-j\omega t_0}$

Modulation: $v(t)e^{j\omega_0 t} \leftrightarrow V(f - f_0)$

Convolution: $\int_{-\infty}^{\infty} v_1(\lambda)v_2(t - \lambda)d\lambda \leftrightarrow V_1(f)V_2(f)$

Multiplication: $v_1(t)v_2(t) \leftrightarrow \int_{-\infty}^{\infty} V_1(\gamma)V_2(f - \gamma)d\gamma$

12.2

12.1.2 Use of Properties

Here we show how to use these properties and how they can simplify your life.

Problem 12.1.1. Use the differentiation property to find the Fourier transform of the square pulse shown in Fig. 12.1.1.

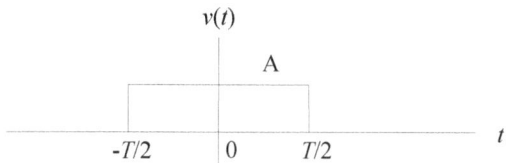

$v(t)$

A

$-T/2$ 0 $T/2$ t

Fig. 12.1.1

Solution: Figure 12.1.2 shows the derivative $v'(t)$. Since the transform of a δ-function occurring at t_0 is given by

$$\delta(t - t_0) \leftrightarrow e^{-j\omega t_0}$$

Then the transform of the two δ-functions in the diagram is

$$v'(t) \leftrightarrow Ae^{\frac{j\omega T}{2}} - Ae^{-\frac{j\omega T}{2}}$$

Now apply the differentiation property of obtain

$$v'(t) \leftrightarrow j\omega V(\omega)$$

Set these two equal and solve for $V(\omega)$ to obtain

$$V(\omega) = \frac{Ae^{\frac{j\omega T}{2}} - Ae^{-\frac{j\omega T}{2}}}{j\omega} = AT\left[\frac{\sin(\omega T/2)}{\omega T/2}\right]$$

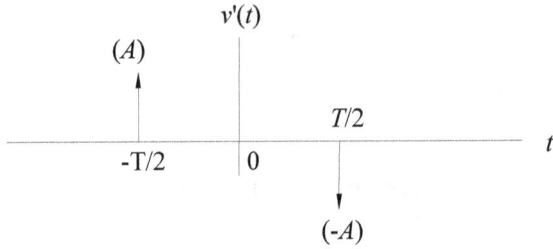

Fig. 12.1.2

Problem 12.1.2. Use the differentiation property to find the Fourier transform of $f(t)$ in Fig. 12.1.3.

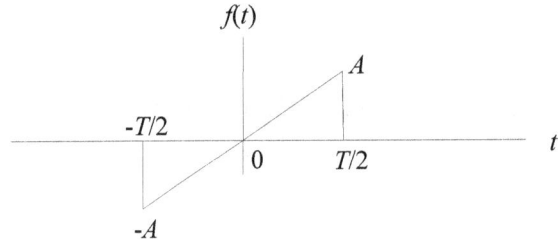

Fig. 12.1.3

Answer: (Hint: You will need to differentiate twice.)

$$F(\omega) = \frac{2A}{j\omega}\left[\frac{sin(\omega T/2)}{\omega T/2} - \cos(\omega T/2)\right]$$

Problem 12.1.3. Use the differentiation property to find the transform of the cosine pulse $f(t)$ in Fig. 12.1.4.

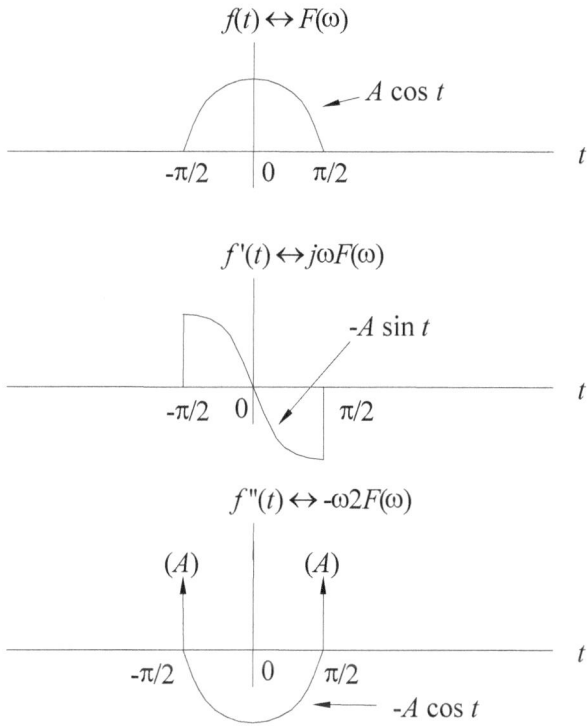

$f(t) \leftrightarrow F(\omega)$

$A \cos t$

$-\pi/2 \quad 0 \quad \pi/2$

t

$f'(t) \leftrightarrow j\omega F(\omega)$

$-A \sin t$

$-\pi/2 \quad 0 \quad \pi/2$

t

$f''(t) \leftrightarrow -\omega2 F(\omega)$

$(A) \qquad (A)$

$-\pi/2 \quad 0 \quad \pi/2$

t

$-A \cos t$

Fig. 12.1.4

Solution: Notice that $\ddot{f}(t)$ would be proportional to $f(t)$ if it were not for the two δ-functions. But these can be subtracted. Define $g(t)$ by:

$$g(t) = \ddot{f}(t) - A\delta(t + \pi/2) - A\delta(t - \pi/2)$$

Therefore the transform of g(t) is given by:

$$G(\omega) = -\omega^2 F(\omega) - Ae^{j\omega\pi/2} - Ae^{-j\omega\pi/2} \tag{12.1.2}$$

But how do we find $F(\omega)$? Since $g(t)$ is related to $f(t)$ by $g(t) = -f(t)$, then $G(\omega)$ must be related to $F(\omega)$ by

$$G(\omega) = -F(\omega)$$

Combining this with Eq. 12.1.2 gives

$$-F(\omega) = -\omega^2 F(\omega) - Ae^{j\omega\pi/2} - Ae^{-j\omega\pi/2}$$

Or

$$F(\omega) = \frac{A}{1 - \omega^2}\left[e^{j\omega\pi/2} + e^{-j\omega\pi/2}\right] = \frac{A}{1 - \omega^2}\cos(\omega\pi/2)$$

Problem 12.1.4. Use the delay property along with the solution to Problem 12.1.1 to find the transform of $g(t)$ in Fig. 12.1.5.

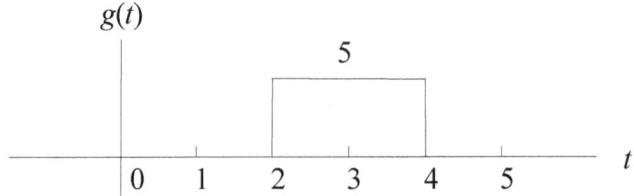

Fig. 12.1.5

Solution: The function $v(t)$ in Fig. 12.1.1 is related to $g(t)$ by g(t) = v(t – 3). Since

$$V(\omega) = AT\frac{\sin(\omega T/2)}{\omega T/2}$$

Apply the delay property to obtain

$$G(\omega) = V(\omega)e^{-j3\omega} = 10\left[\frac{\sin\omega}{\omega}\right]e^{-j3\omega}$$

Problem 12.1.5. Use the convolution property and the solution to Problem 12.1.1 to find the transform of f(t) in Fig. 12.1.6.

12.6

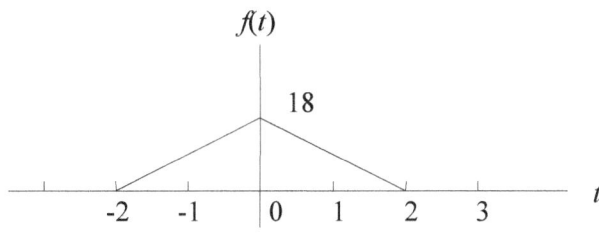

Fig. 12.1.6

Solution: Let $A = 3$ and $T = 2$ in Fig. 12.1.1. Then the function $v(t)$ in Fig. 12.1.1 is related to $f(t)$ by convolution, $f(t) = v(t)*v(t)$. Therefore by the convolution property

$$F(\omega) = [V(\omega)]^2 = 36 \left[\frac{sin\omega}{\omega}\right]^2$$

Problem 12.1.6. Use the differentiation property to solve the previous problem.

12.1.3. The Transform of Power Signals

Although power signals do not satisfy the Dirichlet conditions, the transform of many power signals exist as distribution. This conveniently allows us to combine our notation for the Fourier series and transform. We now set about using the properties to derive the transform of several power signals.

Problem 12.1.7. Find the transform of the eternal exponential signal $g(t) = e^{j\omega_0 t}$

Solution: Before getting into the solution, consider how one would plot this signal versus time. It is complex-valued so it will take two plots, or one three-dimensional plot.

Think of a spiral notebook. The binding on this notebook is a spiral that illustrates how a plot of $g(t)$ looks. Picture a rotating vector (called a phasor) that travels down the time axis as it rotates at unit distance from the axis. This is the plot of $g(t) = e^{j\omega_0 t}$.

Let us take this one step further. Suppose the complex exponential signal $g(t)$ is multiplied by an arbitrary time function. Now the product $v(t)g(t)$ is a phasor that rotates at ω_0 radians per second and travels down the time axis, but changes amplitude according to the dictates of $v(t)$.

One more step, the Fourier transform:

$$V(\omega) = \frac{1}{2\pi} \int_{-\infty}^{\infty} v(t)e^{-j\omega t}\, dt$$

(The phasor here rotates CW instead of CCW, but this is immaterial to our discussion.) Think of this transform as the integral of a modulated phasor over all time. The value of this integral is a complex number that depends on $v(t)$ and ω. For a particular $v(t)$ the value of the integral is a number that changes with ω. It is this function of ω (divided by 2π) that is the Fourier transform of $v(t)$.

Now back to our problem: How to find the transform of $g(t) = e^{j\omega_0 t}$. The transform of a constant $v(t) = 1$ is $\delta(f)$ or $2\pi\delta(\omega)$. This is easily seen by direct application of the inverse Fourier transform.

$$v(t) = \int_{-\omega}^{\infty} \delta(f)e^{j2\pi ft}\, df = 1$$

or

$$v(t) = \frac{1}{2\pi} \int_{-\infty}^{\infty} 2\pi\delta(\omega)e^{j\omega_0 t}\, d\omega = 1$$

Since the eternal exponential $g(t)$ is related to $v(t)$ by

$$g(t) = v(t)e^{j\omega_0 t}$$

Apply the modulation property to obtain

12.8

$$e^{j\omega_0 t} \leftrightarrow \delta(f - f_0) = 2\pi\delta(\omega - \omega_0) \tag{12.1.3}$$

Problem 12.1.8. Find the transform of $f(t) = \sin(\omega_o t)$.

Solution: Since

$$\sin(\omega_0 t) = \frac{1}{2j}e^{j\omega_0 t} - \frac{1}{2j}e^{-j\omega_0 t}$$

Apply the superposition property to Eq. 12.1.3 to obtain

$$\sin(\omega_0 t) \leftrightarrow \frac{1}{2j}\delta(f - f_0) - \frac{1}{2j}\delta(f + f_0) \tag{12.1.4}$$

Problem 12.1.9. Find the transform of the signum function $s(t)$ in Fig. 12.1.7.

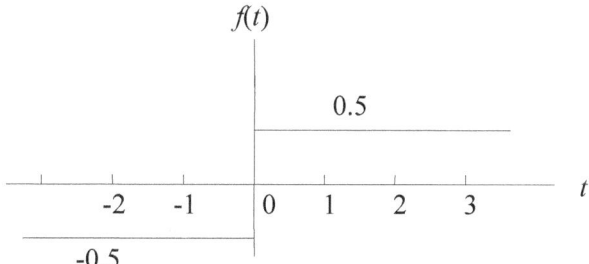

Fig. 12.1.7

Solution: The derivative of $s(t)$ is an impulse of strength 1. Apply the differentiation property to obtain

$$s'(t) = \delta(t) \leftrightarrow j\omega S(f)$$

But the transform of $\delta(t)$ is 1, so $1 = j\omega S(f)$, or

12.9

$$S(f) = \frac{1}{2j}$$ (12.1.5)

Problem 12.1.10. Use the superposition property and Eq. 12.1.5 to find the transform of the unit step $u(t)$ shown in Fig. 12.1.8.

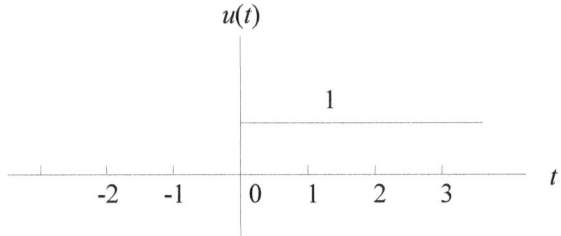

$u(t)$

1

-2   -1   0   1   2   3 t

Fig. 12.1.8

Solution: The unit step $u(t)$ is related to the signum function $s(t)$ by

$$u(t) = s(t) + \frac{1}{2}$$

Hence
$$U(f) = S(f) + \frac{1}{2}\delta(f) = \frac{1}{j\omega} + \frac{1}{2}\delta(f)$$

Note: The derivatives of $u(t)$ and $s(t)$ are equal. We cannot apply the differentiation property directly to find the transform of $u(t)$ since the dc level is lost in taking the derivative.

We can generalize this observation. If you take the derivative of a power signal, the dc level (zero frequency component) is lost. This does not apply to energy signals, however, because an energy signal has a dc level of zero.

Problem 12.1.11. Find the transform of $p(t) = [\sin \omega_0 t] u(t)$ shown in Fig. 12.1.9.

Solution: Apply the multiplication property. Since $p(t)$ is the product of $f(t)$ in Eq. 12.1.4 with $u(t)$ in Eq. 12.1.6, we can convolve $F(f)$ with $U(f)$ to obtain

$$P(f) = F(f) * U(f) = \frac{1}{4j}\delta(f - f_0) - \frac{1}{4j}\delta(f + f_0) - \frac{\omega_0}{\omega^2 - \omega_0^2} \qquad (12.1.7)$$

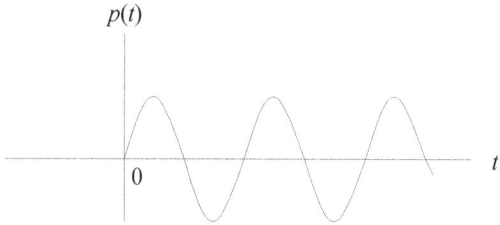

$p(t)$

0

t

Fig. 12.1.9

Problem 12.1.12. Find the transform of $q(t) = \left[\cos\omega_0 t\right] u(t)$.

Answer: $\qquad Q(f) = \frac{1}{4}\delta(f - f_0) + \frac{1}{4}\delta(f + f_0) - \frac{j\omega_0}{\omega^2 - \omega_0^2} \qquad (12.1.8)$

Self Test, Objective 12.1. Find the transform of the functions in Fig. 12.1.10 below.

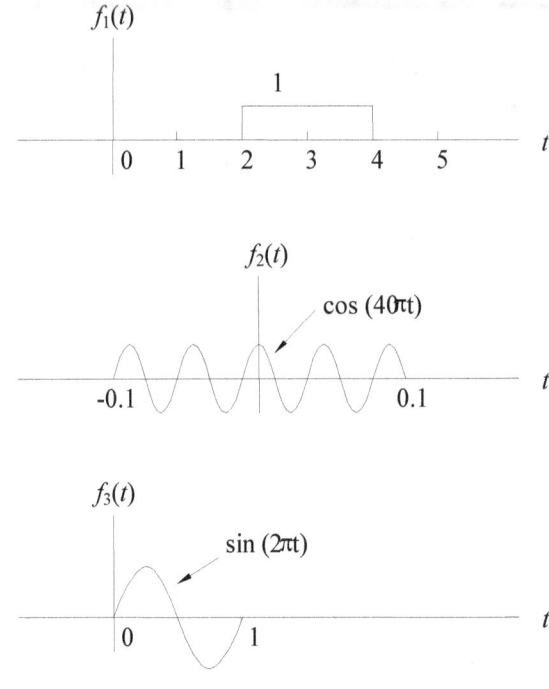

Fig. 12.1.10

Objective 12.2. Find bounds on the spectrum of a given time function by using the concepts of content, variation, and wiggliness.

in addition to its use in evaluating the series and transform of a signal, the differentiation property has another use. We may derive bounds on the magnitude of the spectrum and relate these bounds to the time function. This might be useful in situations where the task of finding the frequency spectrum is tedious and only the approximate spectrum is needed. (The term "frequency spectrum" refers to the coefficients $V(f_k)$ if $v(t)$ is periodic, and to the function $V(f)$ if $v(t)$ is an energy signal.)

12.12

First suppose $v(t)$ is periodic. The content, variation, and wiggliness of $v(t)$ are defined as

$$content_p = \frac{1}{T} \int_{t_1}^{t_1+T} |v(t)| \, dt \tag{12.2.1}$$

$$variation_p = \frac{1}{T} \int_{t_1}^{t_1+T} \left| \frac{dv}{dt} \right| \, dt \tag{12.2.2}$$

$$wiggliness_p = \frac{1}{T} \int_{t_1}^{t_1+T} \left| \frac{d^2v}{dt^2} \right| \, dt \tag{12.2.3}$$

The frequency spectrum is related to the derivatives by

$$V_k = \frac{1}{T} \int_{t_1}^{t_1+T} v(t) e^{-j\omega_k t} \, dt \tag{12.2.4}$$

$$j\omega_k V_k = \frac{1}{T} \int_{t_1}^{t_1+T} \frac{dv}{dt} e^{-j\omega_k t} \, dt \tag{12.2.5}$$

$$(j\omega_k)^2 V_k = \frac{1}{T} \int_{t_1}^{t_1+T} \frac{d^2v}{dt^2} e^{-j\omega_k t} \, dt \tag{12.2.6}$$

Now to derive the bounds take absolute values of both sides of these last three equations and use the relationship

$$\left| \int_a^b f(t) g(t) \, dt \right| \le \int_a^b |f(t)| |g(t)| \, dt \tag{12.2.7}$$

to obtain

$$|V_k| \le \frac{1}{T} \int_{t_1}^{t_1+T} |v(t)| \, dt = content_p \tag{12.2.8}$$

$$|\omega_k| |V_k| \le \frac{1}{T} \int_{t_1}^{t_1+T} \left| \frac{dv}{dt} \right| \, dt = variation_p \tag{12.2.9}$$

$$\omega_k^2 |V_k| \le \frac{1}{T} \int_{t_1}^{t_1+T} \left| \frac{d^2v}{dt^2} \right| \, dt = wiggliness_p \tag{12.2.10}$$

This leads to the bounds, given by

12.13

$$|V_k| \le \begin{cases} content_p \\ variation_p / |\omega_k| \\ wiggliness_p / \omega_k^2 \end{cases} \qquad (12.2.11)$$

Next, the content, variation, and wiggliness for aperiodic energy signals is defined as

$$content_a = \int_{-\infty}^{\infty} |v(t)| dt \qquad (12.2.12)$$

$$variation_a = \int_{-\infty}^{\infty} \left| \frac{dv}{dt} \right| dt \qquad (12.2.13)$$

$$wiggliness_a = \int_{-\infty}^{\infty} \left| \frac{d^2 v}{dt^2} \right| dt \qquad (12.2.14)$$

Note that the subscript p is used for periodic signals, and a is used for aperiodic energy signals.

Similar reasoning leads to the bounds for energy signals.

$$|V(f)| \le \begin{cases} content_a \\ variation_a / |\omega| \\ wiggliness_a / \omega^2 \end{cases} \qquad (12.2.15)$$

Here is a less formal view of content, variation, and wiggliness: The content is the total area (both positive and negative) under the curve of the function. The variation is the total change in altitude of the curve, that is, the total vertical distance traveled as you trace the curve. The wiggliness, which is surely the most descriptive of these terms, is a measure of how much the curve wiggles, that is, the total change in the slope of the curve.

This description is for energy signals. For periodic power signals we must divide by the period. Thus content, variation, and wiggliness for periodic signals depends on the period.

For example, an energy signal $v(t)$ and its derivatives are shown in Fig. 12.2.1. Notice that you could determine the content, variation, and wiggliness from the plot of $v(t)$ alone. The content is the total area under $v(t)$, and that is 3.

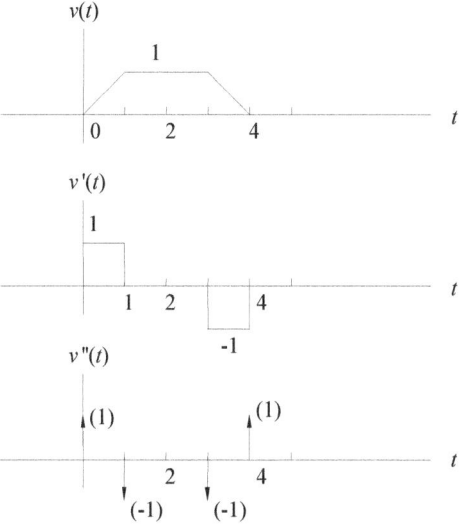

Fig. 12.2.1

The variation is the total up and down distance as the curve is traced. It goes up from 0 to 1, then goes down from 1 to 0, so the total vertical distance is 2. Thus the variation is 2.

The slope changes at the origin from 0 to 1, then at $t = 1$ it changes from 1 to 0. At $t = 3$ it changes from 0 to -1, and again at $t = 4$ from -1 to 0. Thus the total change in slope is 4, and this is the wiggliness.

According to Eq. 12.2.15, the spectrum for $v(t)$ in Fig. 12.2.1 is bounded by

$$|V(f)| \leq \begin{cases} 3, \\ 2/|\omega| \\ 4/\omega^2 \end{cases}$$

12.15

Figure 12.2.2 shows these bounds with the magnitude of $V(f)$.

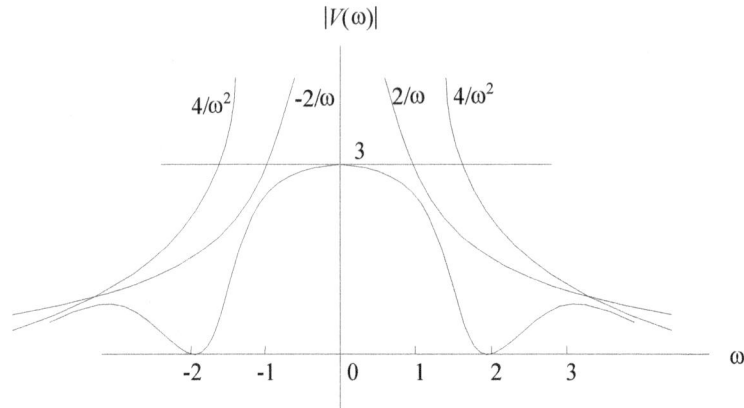

Fig. 12.2.2

Problem 12.2.1. Bound the spectrum for the periodic waveform shown in Fig. 12.2.3.

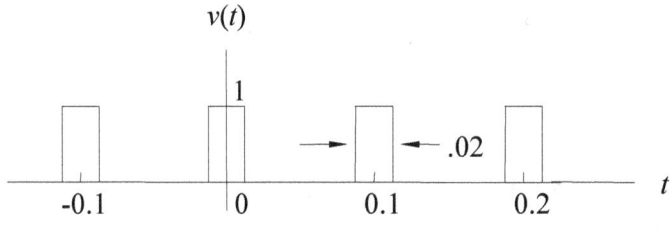

Fig. 12.2.3

Solution: Use Eqs. 12.2.8 through 12.2.10 to calculate the content, variation, and wiggliness.

12.16

$$content_p = \tfrac{1}{0.1} \int_{-.01}^{.01} 1 dt = 0.2$$

$$variation_p = \tfrac{1}{0.1} \int_{-0.05}^{0.05} [\delta(t+0.01) + \delta(t-0.01)] dt = 20$$

$$wiggliness_p = \infty$$

This gives bounds

$$|V_k| \leq \begin{cases} 0.2 \\ 20 \\ \dfrac{|\omega_k|}{\infty} \end{cases}$$

Figure 12.2.4 shows the magnitude spectrum with these bounds superimposed, where the spectrum is given by

$$V_k = 0.2 \frac{\sin 0.2k\pi}{0.2k\pi}$$

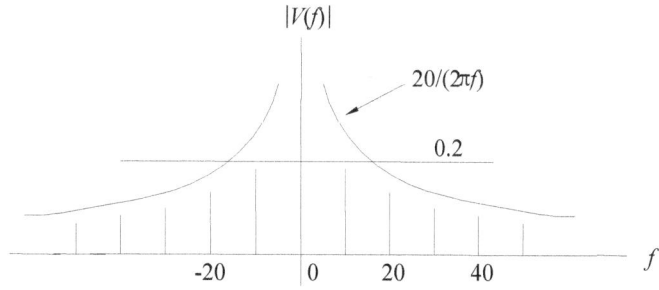

Fig. 12.2.4

12.17

Problem 12.2.2. Bound the spectrum for the aperiodic signal shown in Fig. 12.2.5.

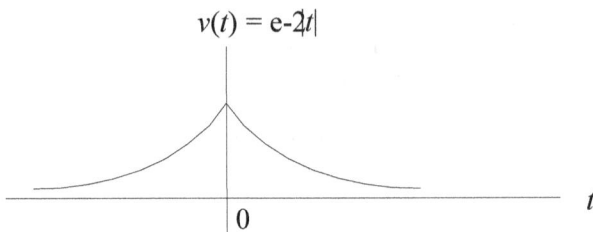

$$v(t) = e-2t|$$

Fig. 12.2.5

Solution: The first and second derivatives of $v(t)$ are shown in Fig. 12.2.6. Using Eqs. 12.2.12 through 12.2.14 gives

$$content_a = 2\int_0^\infty e^{-2t} dt = 1$$

$$variation_a = 2\int_0^\infty 2e^{-2t} dt = 2$$

$$wiggliness_a = 2\int_0^\infty 4e^{-2t} dt + \int_{-\infty}^\infty 4\delta(t) dt = 8$$

(We have taken advantage of the symmetry in each diagram to reduce the number of integrals.)

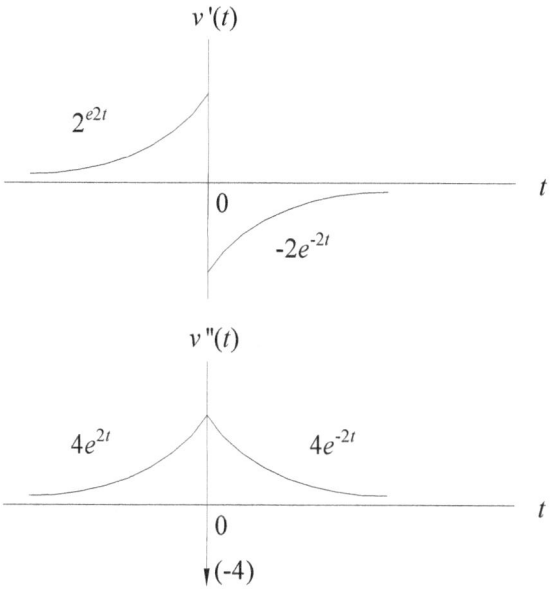

Fig. 12.2.6

Self Test, Objective 12.2.

Find bounds on the spectrum for each function in Figs. 12.2.7 and 12.2.8.

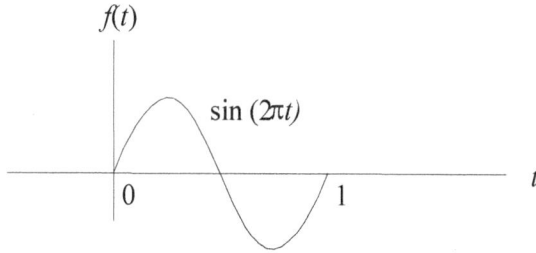

Fig. 12.2.7

12.19

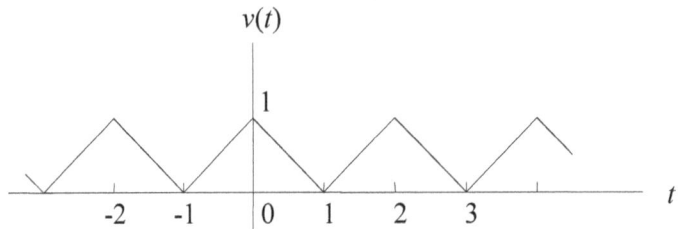

Fig. 12.2.8

Objective 12.3. Use the Paley-Wiener theorem to test the magnitude of transfer functions for physical realizability.

The term "physical realizability" seems to imply that we can actually construct the system. This is unfortunate, because in system theory the term has a precise meaning, causalty. That is, the system output must not precede the input. Figure 12.3.1 illustrates the impulse response of two systems. The first is not physically realizable because the output would occur before the input is applied. This system is called anticipatory. (Note that anticipatory systems can be simulated. A digital computer can store and operate on data in an anticipatory manner.) The system with impulse response $h_2(t)$ in Fig. 12.3.1 is causal, and is therefore called physically realizable.

Note: It would be difficult to use the criterion "can it be built" to decide if a system is physically realizable. New and better ways to build systems are continually discovered.

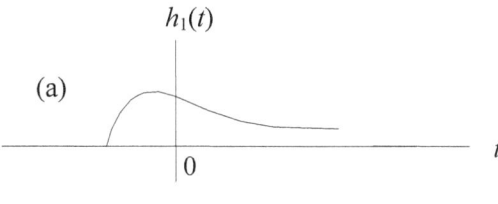

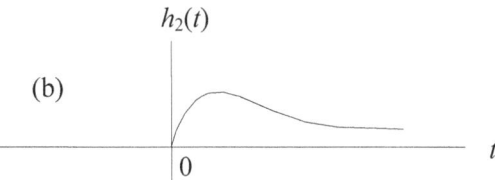

Fig. 12.3.1

The Paley-Wiener Theorem

You can determine at a glance if a system is physically realizable from the impulse response. But in the frequency domain how can $H(\omega)$ provide this information? The Paley-Wiener criterion partially answers this question and provides valuable insight into the problem.

If the magnitude of the transfer function is square integrable,

$$\int_{-\infty}^{\infty} |H(\omega)|^2 \, d\omega < \infty \qquad (12.3.1)$$

then we may apply the following (Paley-Wiener) criterion. A necessary and sufficient condition for $|H(\omega)|$ to be physically realizable is that

$$\int_{-\infty}^{\infty} \frac{|ln|H(\omega)||}{1+\omega^2} \, d\omega < \infty \qquad (12.3.2)$$

Note: This does not say that $H(\omega)$ is automatically causal if it satisfies the Paley-Wiener criterion. An appropriate phase function must go with the magnitude

12.21

function so that the combination is causal. This criterion does, however, say that if $|H(\omega)|$ satisfies both Eq. 12.3.1 and 12.3.2 then it is possible to select such a phase function.

Here are some important conclusions from the Paley-Wiener criterion:

1. The ideal filter, so often used in examples, cannot be physically realizable.

2. A waveform cannot be limited in both time and frequency. (Does this sound like the uncertainty principle?)

Let us apply the Paley-Wiener criterion to the ideal filter, pictured in Fig. 12.3.2. The magnitude function is square integrable, but it does not satisfy the relation in Eq. 12.3.2. To see this, consider the frequencies greater than f_m. Here

$$H(\omega) = 0, \quad \ln H(\omega) = -\infty, \quad \text{and} \quad \left|\ln H(\omega)\right| = \infty$$

The area under this curve is infinite, and Eq. 12.3.2 is not satisfied.

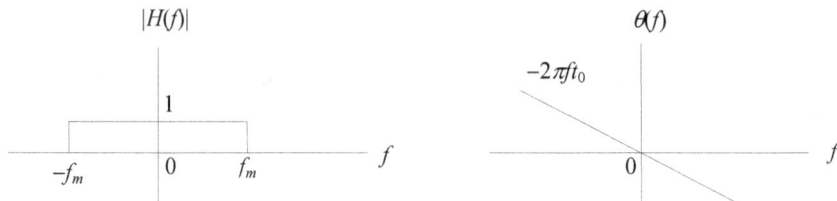

Fig. 12.3.2

To better understand why this filter is not causal, let us find the impulse response. Using the particular phase function in Fig. 12.3.2 gives

$$H(f) = \begin{cases} e^{-j2\pi t_0 f}, & -f_m < f < f_m \\ 0, & otherwise \end{cases}$$

Therefore

$$h(t) = \int_{-\infty}^{\infty} H(f)e^{j2\pi ft} df = 2f_m \frac{\sin(2\pi f_m(t-t_0))}{2\pi f_m(t-t_0)}$$

This is plotted in Fig. 12.3.3. Notice that $h(t)$ is not zero over any interval of time, and regardless of how big t_0 is chosen, the system is anticipatory.

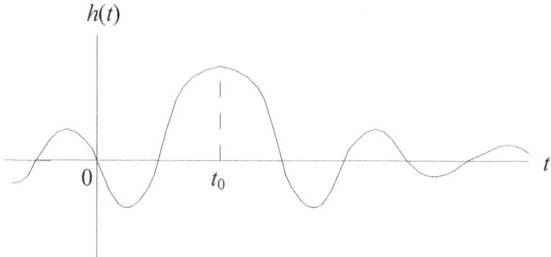

Fig. 12.3.3

Notes: a) The particular phase function chosen here is linear. You may ask if any other phase function could possibly permit $h(t)$ to be causal. The Paley-Wiener theorem assures us that this is not possible.

b) The Paley-Wiener theorem shows that if a function is zero over any interval then its transform cannot be zero over any interval.

Problem 12.3.1. Which of the magnitude functions in Fig. 12.3.4 satisfy the Paley-Wiener theorem? i.e., which could be physically realizable?

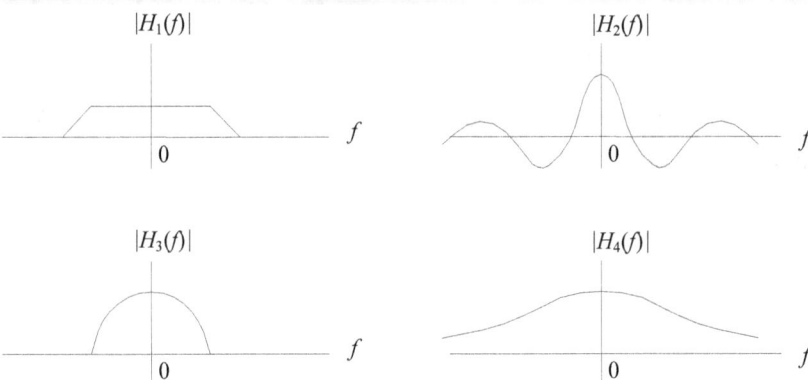

Fig. 12.3.4

Answer: The even numbered functions.

Problem 12.3.2. Select a suitable phase function so that H_2 and H_4 in Fig. 12.3.4 will be causal, where

$$|H_2(\omega)| = \frac{sin\omega}{\omega} \qquad\qquad |H_4(\omega)| = \left|\frac{1}{1+j\omega}\right|$$

Problem 12.3.3. Select a suitable phase function so that H_2 and H_4 will <u>not</u> be causal.

Self Test, Objective 12.3.

If you understand the above three problems then you have completed Objective 12.3.

Self-Test Answrs

<u>Objective 12.1.</u>

$$F(\omega) = \frac{e^{-j2\omega} - e^{-j4\omega}}{j\omega}$$

<u>Objective 12.2.</u>

For Fig. 12.2.7

$$|V(\omega)| \leq \begin{cases} 2/\pi \\ 4/|\omega| \\ 2/\pi\omega^2 \end{cases}$$

For Fig. 12.2.8

$$|V_k| \leq \begin{cases} 1/2 \\ 1/2|\omega_k| \\ 2/\omega_k^2 \end{cases}$$

12.25

Chapter 13.

Laplace Transform

> Objectives: After completing this chapter you should be able to do the following:
>
> 13.1. Determine (select) which functions have a Laplace transform.
>
> 13.2. Find the Laplace transform for a given time function.
>
> 13.3. Select the time function corresponding to a given Laplace transform.

Rationale: The Fourier transform is too restrictive. We were able to express the transfer function for an LTI system as a function of the complex variable $s = \sigma + j\omega$, but in using the Fourier transform we always had $s = j\omega$. The Laplace transform is a generalization of the Fourier transform where frequency is no longer restricted to the line $\sigma = 0$ in the complex plane. There are several advantages to this generalization, but the principle advantage is that we can find the transform of many more functions. Since the transfer function can be expressed as a function of s, this will allow us to solve a larger class of problems.

Objective 13.1. Determine (select) which functions have a Laplace transform.

33.1.1 Bilateral Laplace Transform

There are two forms of the Fourier transform, the bilateral or two-sided, and the unilateral or one-sided form. The two-sided form is given by

$$V(s) = \int_{-\infty}^{\infty} v(t)e^{-st}\,dt \tag{13.1.1}$$

The one-sided version is identical to this except the lower limit on the integral is zero instead of minus infinity.

Figure 13.1.1 shows the Laplace transform pictured as an operator. Supply a suitable function $v(t)$ to the black box and it inserts $v(t)$ in the integral in Eq. 13.1.1, turns the crank, and then spits out $V(s)$.

Fig. 13.1.1

In the Fourier transform the result (output) of the operator was a function $V(f)$ where f was a real variable. The values of V were complex, but the values of f were real. Now the variable s is also allowed to be complex, written $s = \sigma + j\omega$.

The inverse Laplace transform is given by

$$v(t) = \frac{1}{2\pi j} \int_{\sigma - j\omega}^{\sigma + j\omega} V(s)e^{st}ds \qquad (13.1.2)$$

This operator is pictured in Fig. 13.1.2. Supply a suitable function $V(s)$ to the black box and it inserts $V(s)$ in the integral in Eq. 13.1.2, turns the crank, and then spits out $v(t)$.

Fig. 13.1.2

13.2

Since s is a complex variable the limits of integration are complex. Otherwise, the Fourier and Laplace transforms are quite similar. In fact, the Laplace transform can be derived from the Fourier transform by substitution of variable $s = \sigma + j\omega$. We will not do so here because it would add little to this objective.

13.1.2. Unilateral Laplace Transform

The unilateral or one-sided form is widely used in circuit theory because most time functions start at $t = 0$. This form is given by

$$V(s) = \int_0^\infty v(t) e^{-st} dt$$

The inverse transform remains unchanged, given by Eq. 13.1.2.

13.1.3. Which Functions Have a Laplace Transform?

Look at Eq. 13.1.1. The magnitude of $V(s)$ must be finite if $v(t)$ has a transform. This is written as

$$|V(s)| = \left| \int_{-\infty}^\infty v(t) e^{-st} dt \right| < +\infty \qquad (13.1.4)$$

But since

$$|v(t) e^{-st}| = |v(t)| e^{-\sigma t}$$

the existence of the Laplace transform is guaranteed if

$$\int_{-\infty}^\infty |v(t)| e^{-\sigma t} dt < +\infty \qquad (13.1.5)$$

Obviously the value of $|V(s)|$ in Eq. 13.1.4 depends on the value of σ used in Eq. 13.1.5. Equation 13.1.4 might be satisfied for some values of σ but not for others. If we can find a value of σ for which $|V(s)|$ is finite, then the Laplace transform exists for that value of σ. Therefore we are faced with the following problem: Given a function $v(t)$, how do we either 1) determine that $v(t)$ does not have a transform, or 2) how do we choose the right value of σ to satisfy Eq. 13.1.5?

If there exists a positive finite number M so that for real α and β we have

$$|v(t)| \le \begin{cases} Me^{\alpha t} & for\ t > 0 \\ Me^{\beta t} & for\ t < 0 \end{cases} \tag{13.1.6}$$

then Eq. 13.1.5 is satisfied for any value of σ between α and β. That is, for

$$\alpha < \sigma < \beta \tag{13.1.7}$$

In order to show that Eqs. 13.1.6 and 7 define the domain of the Laplace transform, substitute into Eq. 13.1.4 to obtain

$$|V(s)| \le \left| \int_{-\infty}^{0} Me^{(\beta-s)t} dt + \int_{0}^{\infty} Me^{(\alpha-s)t} dt \right|$$

$$= \left| \frac{M}{\beta - s} e^{(\beta-s)t} \Big|_{-\infty}^{0} + \frac{M}{\alpha - s} e^{(\alpha-s)t} \Big|_{0}^{\infty} \right|$$

It is evident that each term will be finite so long as σ satisfies Eq. 13.1.6.

Notes: a) A positive-time function is one that satisfies the condition $f(t) = 0$, $t < 0$. A negative-time function satisfies the condition $f(t) = 0$, $t > 0$. Now consider Eqs. 13.1.6 and 7. It is evident that the region of convergence for a positive time function is to the right of some boundary in the complex plane, i.e., $\alpha < \sigma$. For a negative time function, the region of convergence is to the left of β, i.e., $\sigma < \beta$. These regions of convergence are pictured in Fig. 13.1.3.

b) Notice our original condition for convergence given by Eq. 13.1.4, namely $|V(s)| < +\infty$. This means that if the transform of a positive-time function has any poles (values of s for which $V(s) = \infty$) they must lie to the left of $\sigma = \alpha$ in the complex plane. Likewise, any poles of a negative-time function must lie to the right of the $\sigma = \beta$ line.

c) If the region of convergence includes the $\sigma = 0$ line, then $v(t)$ has a Fourier transform, and it is found from the bilateral Laplace transform by setting $\sigma = 0$, or $s = j\omega$.

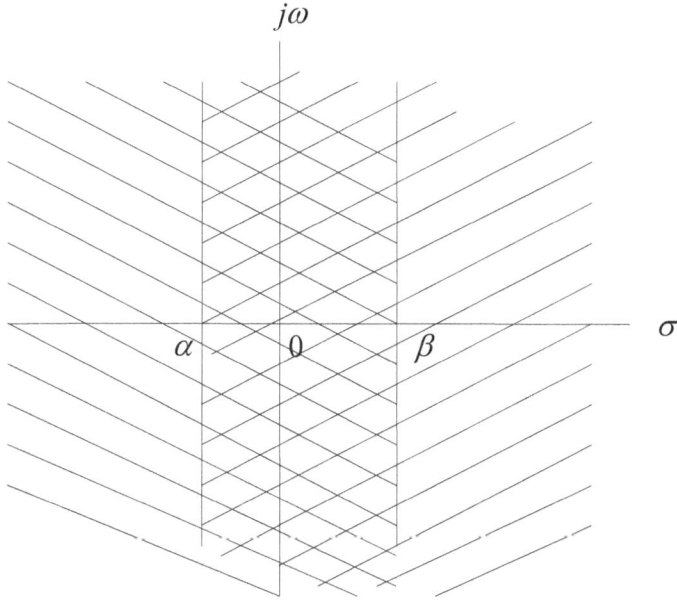

Fig. 13.1.3

We will now find the Laplace transform for several functions along with the region of convergence in the s-plane.

13.5

Problem 13.1.1. Find the Laplace transform and region of convergence for the unit step function. (Fig. 13.1.4.)

Solution: If $v(t) = u(t)$ then the Laplace transform is given by

$$V(s) = \int_0^\infty e^{-st} dt = \frac{1}{s} \qquad (13.1.8)$$

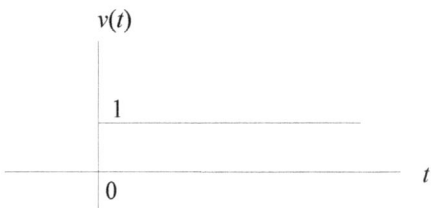

$v(t)$

1

0

t

Fig. .13.1.4

Use Eq. 13.1.6 to find the region of convergence. Set $M = 2$ (say) so that for $t > 0$, any value of $\alpha > 0$ will satisfy the condition

$$Me^{\alpha t} \ge |v(t)|, \ t > 0 \qquad (13..1.9)$$

Thus the minimum value of α is zero. Any value of σ greater than this minimum α is in the region of convergence for the positive-time function.

Note: The region of convergence does not include $\sigma = 0$. Therefore the Fourier transform of a unit step cannot be found by simply setting $s = j\omega$ in Eq. 13.1.8.

Problem 13.1.2. Find the Laplace transform and region of convergence for the two-sided exponential function in Fig. 13.1.5. The function $v(t)$ is given by

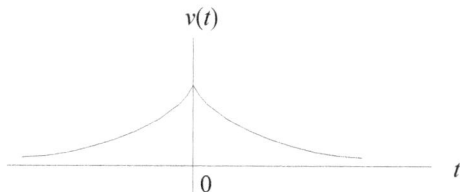

$v(t)$

Fig. 13.1.5

$$v(t) = \begin{cases} e^{at}, & t < 0 \\ e^{-at}, & t > 0 \end{cases}$$

Solution:

$$V(s) = \int_{-\infty}^{0} e^{at} e^{-st}\, dt + \int_{0}^{\infty} e^{-at} e^{-st}\, dt = \frac{1}{a-s} + \frac{1}{a+s} = \frac{-2a}{s^2-a^2} \qquad (13.1.10)$$

The region of convergence is given by $-a < \sigma < a$. To see this, first consider $t > 0$. For any $M > 1$ in Eq. 13.1.6 we have

$$Me^{\alpha t} > e^{-at}, \quad t > 0$$

so long as $\alpha \geq -a$. This condition will not be true for any $\alpha < -a$ for large values of t, no matter how large M is chosen.

Next, consider $t < 0$. For any $M > 1$ in Eq. 13.1.6 we have

$$Me^{\beta t} > e^{at}, \quad t < 0$$

so long as $\beta \leq a$.

Note: Now the region of convergence does include $\sigma = 0$. Therefore the Fourier transform is, from Eq. 13.1.10, given by setting $s = j\omega$ to obtain

$$V(\omega) = \frac{2a}{\omega^2 + a^2}$$

Problem 13.1.3. Find the Laplace transform and region of convergence for the eternal sinusoid if Fig. 13.1.6.

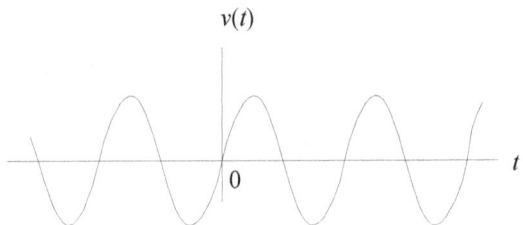

$v(t)$

0

t

Fig. 13.1.6

Solution: Testing this function in Eq. 13.1.6 shows that $\sigma > 0$ is the region of convergence for $t > 0$. For $t < 0$ the region of convergence is $\sigma < 0$. Thus there is no value of σ for which the transform converges and this function does not have a Laplace transform.

But recall that in Chapter 12 the Fourier transform of $v(t)$ was given by two δ-functions.

$$V(f) = \tfrac{1}{2j}\delta(f - f_0) - \tfrac{1}{2j}\delta(f + f_0)$$

It seems that if $\sigma = 0$ is the boundary of the convergence region then the Fourier transform exists as distribution (that is, δ-functions are part of the transform.)

Problem 13.1.3. Find the Laplace transform and region of convergence for the positive-time sinusoid $v(t) = \sin(\omega_0 t)u(t)$ in Fig. 13.1.7.

13.8

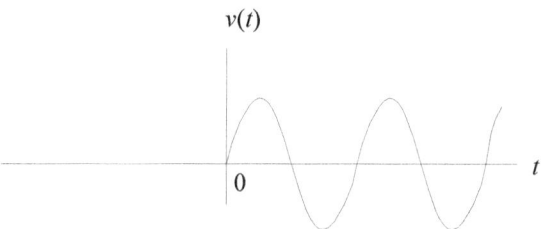

Fig. 13.1.7

Solution:

$$V(s) = \int_0^\infty \sin(\omega_0 t)e^{-st}dt = \frac{\omega_0}{s^2+\omega_0^2} \qquad (13.1.11)$$

The region of convergence is $\sigma > 0$.

Note: It will not do to substitute $s = j\omega$ into Eq. 13.1.11 to find the Fourier transform since the $j\omega$ axis is on the boundary. But the Fourier transform exists as distribution, given by

$$V(\omega) = \frac{\pi}{2j}[\delta(\omega - \omega_0) - \delta(\omega + \omega_0)] + \frac{\omega_0}{\omega_0^2 - \omega^2}$$

Self Test, Objective 13.1. Determine the region of convergence for the following functions.

1. $e^{2t}u(t)$ 3. $e^{-3|t|}$

2. $\sin(2\pi t)$ 4. $e^{2t}u(-t)$

Objective 13.2. Find the Laplace transform for a given time function.

This is just a matter of applying the definition given by Eq. 13.1.1, or, for positive-time functions, Eq. 13.1.3. The properties will prove useful to us and they are identical to those for the Fourier transform with some exceptions. One notable exception is the differentiation property for the single-sided transform. It is given by

$$\frac{dv}{dt} \leftrightarrow sV(s) - v(0) \tag{13.2.1}$$

It is this property that makes the single-sided transform so useful in the solution of differential equations, because the initial conditions are included as a matter of course.

The proof of this property is straightforward. By definition the Laplace transform of the derivative is given by

$$L\left[\frac{dv}{dt}\right] = \int_0^\infty \frac{dv}{dt} e^{-st} dt$$

Integrating by parts gives

$$L\left[\frac{dv}{dt}\right] = v(t)e^{-st}\Big|_0^\infty + s\int_0^\infty v(t)e^{-st} dt$$

For higher derivatives this property is found, either by repeated application of Eq. 13.2.1 or by repeated integration, to be given by

$$\frac{d^n v}{dt} \leftrightarrow s^n V(s) - s^{n-1}v(0) - s^{n-2}\frac{dv(0)}{dt} - \cdots - \frac{d^{n-1}v(0)}{dt^{n-1}} \tag{13.2.2}$$

For completeness here is a list of the more useful properties. These properties apply to either the bilateral or unilateral transform.

Table 13.2.1. Properties of the Laplace Transform

Uniqueness: $v(t) \leftrightarrow V(s)$

Superposition: $av(t) + bw(t) \leftrightarrow aV(s) + bW(s)$

Scaling: $v(t/a) \leftrightarrow aV(as)$ for real $a > 0$

Delay: $v(t - t_0) \leftrightarrow V(s)e^{-st_0}$

Modulation: $v(t)e^{s_0 t} \leftrightarrow V(s - s_0)$

Convolution: $\displaystyle\int_{-\infty}^{\infty} v_1(\lambda)v_2(t - \lambda)d\lambda \leftrightarrow V_1(s)V_2(s)$

Multiplication: $v_1(t)v_2(t) \leftrightarrow \frac{1}{2\pi j}\displaystyle\int_{\sigma-j\infty}^{\sigma+j\infty} V_1(\lambda)V_2(s - \lambda)d\lambda$

Problem 13.2.1. Find the Laplace transform of the pulse shown in Fig. 13.2.1.

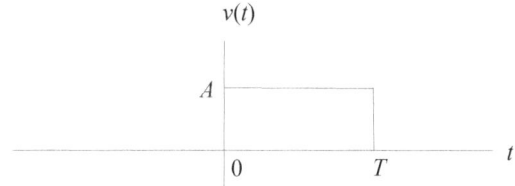

Fig. 13.2.1

Solution: Instead of using the definition let us illustrate the use of properties in finding transforms. First express $v(t)$ as the sum of two step functions.

$$v(t) = Au(t) - Au(t - T)$$

13.11

The delay property applied to the transform of $u(t)$ given in Eq. 13.1.9 gives

$$u(t-T) \leftrightarrow \frac{1}{s} e^{-sT}$$

The superposition property now gives

$$V(s) = \frac{A}{s} - \frac{A}{s} e^{-sT}$$

The region of convergence is the entire s-plane.

Note: The region of convergence for any signal that lasts for a finite time is the entire s-plane.

Problem 13.2.2. Find the Laplace transform of the signal $v(t)$ in Fig. 13.2.2.

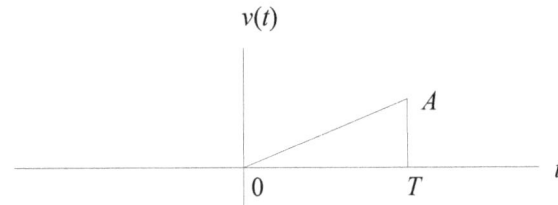

Fig. 13.2.2

Solution: The Laplace transform of a ramp signal $t\,u(t)$ is given by

$$tu(t) \leftrightarrow \frac{1}{s^2}, \quad \sigma > 0$$

The signal in Fig. 13.2.2 can be expressed as the sum of step and ramp signals as shown in Fig. 13.2.3. Therefore

$$V(s) = \frac{A}{T} \left[\frac{1}{s^2} - \frac{1}{s^2} e^{-sT} \right] - \frac{A}{T} e^{-sT}, \quad -\infty < \sigma < \infty$$

13.12

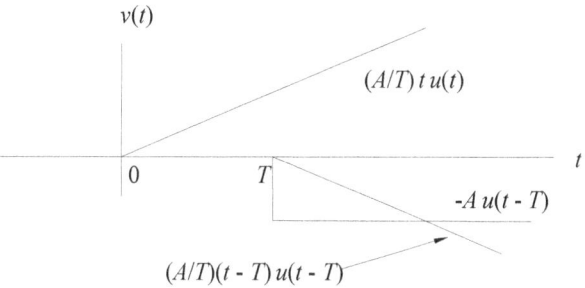

Fig. 13.2.3

Notes: a) You can apply the definition in Eq. 13.1.1 directly to get the same answer.

b) The region of convergence is the entire s-plane even though the region of convergence for each ramp and step is only the half-plane.

c) Since the $\sigma = 0$ line is in the region of convergence you can find the Fourier transform directly by substituting $s = j\omega$ into the expression for $V(s)$.

Self Test, Objective 13.2. Find the Laplace transform of the following functions. Include the region of convergence.

$v_1(t) = (t-1)u(t)$ $\qquad$ $v_2(t) = (t-)u(t-1)$

$v_3(t) = e^{-a|t|}$ $\qquad$ $v_4(t) = tu(-t)$

Objective 13.3. Select the time function corresponding to a given Laplace transform.

The same frequency function $F(s)$ can correspond to two different time functions if the regions of convergence are different. In this objective we will learn to select the proper time function for a given region of convergence in the s-plane.

13.13

CHAPTER 13

Problem 13.3.1. Find the Laplace transform of $f_1(t)$ and $f_2(t)$ in Fig. 13.3.1.

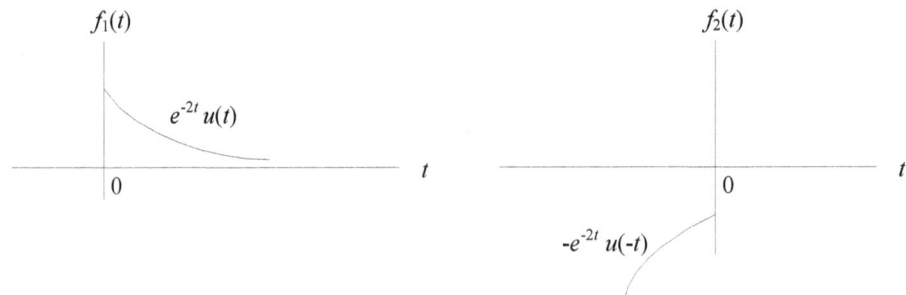

Fig. 13.3.1

Solution:

$$F_1(s) = \int_0^\infty e^{-2t}e^{-st}dt = \frac{1}{s+2}, \quad \sigma > -2$$

$$F_2(s) = \int_{-\infty}^0 -e^{-2t}e^{-st}dt = \frac{1}{s+2}, \quad \sigma < -2$$

Notice that $F_1(s)$ and $F_2(s)$ are the same except for the regions of convergence.

Problem 13.3.2. Find the inverse transform of

$$F(s) = \frac{1}{s+1}, \quad \sigma < -1$$

Answer: $f(t) = -e^{-t}u(-t)$

Problem 13.3.3. Find the inverse transform of

13.14

$$F(s) = \frac{1}{s+1} + \frac{1}{s+2}, \quad -2 < \sigma < -1$$

Solution: Combining the results of problems 13.3.1 and 2 gives

$$f(t) = e^{-2t}u(t) - e^{-t}u(-t)$$

Problem 13.3.4. Find the inverse transform of

$$F(s) = \frac{1}{s+1} + \frac{1}{s+2}, \quad \sigma > -1$$

Answer: $f(t) = e^{-2t}u(t) + e^{-t}u(t)$

Self Test, Objective 13.3. Find the inverse transform of the following functions.

$$F_1(s) = \frac{1}{s+a}, \quad \sigma < -a$$

$$F_2(s) = \frac{1}{s+1}, \quad \sigma < -2$$

$$F_3(s) = \frac{1}{s+a}, \quad \sigma > -a$$

$$F_4(s) = \frac{1}{s}, \quad \sigma > 0$$

13.15

CHAPTER 13

Self Test Answers

Objective 13.1:

1. $\sigma > 2$

2. none

3. $-3 < \sigma < 3$

4. $\sigma < 2$

Objective 13.2:

$$V_1(s) = \frac{1}{s^2} - \frac{1}{s}, \quad \sigma > 0$$

$$V_2(s) = \frac{1}{s^2} e^{-s}, \quad \sigma > 0$$

$$V_3(s) = \frac{-2a}{s^2 - a^2}, \quad -a < \sigma < a$$

$$V_4(s) = \frac{-1}{s^2}, \quad \sigma > 0$$

Objective 13.3:

$$f_1(t) = -e^{-at}u(-t)$$
$$f_2(t) = (-e^{-2t})u(-t)$$
$$f_3(t) = e^{-at}u(t)$$
$$f_4(t) = u(t)$$

Table of common Laplace transform pairs

1. $\delta(t) \leftrightarrow 1, \quad -\infty < \sigma < \infty$

2. $u(t) \leftrightarrow \frac{1}{s}, \quad \sigma > 0$

3. $u(-t) \leftrightarrow \frac{1}{s}, \quad \sigma < 0$

4. $tu(t) \leftrightarrow \frac{1}{s^2}, \quad \sigma > 0$

5. $-tu(-t) \leftrightarrow \frac{1}{s^2}, \quad \sigma < 0$

6. $t^n u(t) \leftrightarrow \frac{n!}{s^{n+1}}, \quad \sigma > 0$

7. $e^{at} u(t) \leftrightarrow \frac{1}{s+a}, \quad \sigma > Re(a)$

8. $-e^{at} u(-t) \leftrightarrow \frac{1}{s+a}, \quad \sigma < Re(a)$

9. $t^n e^{at} u(t) \leftrightarrow \frac{n!}{(s-a)^{n+1}}, \quad \sigma > Re(a)$

10. $cox(at)u(t) \leftrightarrow \frac{s}{s^2+a^2}, \quad \sigma > 0$

11. $\sin(at)\, u(t) \leftrightarrow \frac{a}{s^2+a^2}, \quad \sigma > 0$

13.17

Chapter 14

Response of LTI Systems by Laplace Transform

Objectives: After completing this chapter you should be able to do the following:

14.1. Use partial fraction expansion to express the ratio of polynomials as the sum of partial fractions.
14.2. Find the response of an LTI system to an input signal that is expressed by its Laplace transform.

Rationale: This is a continuation of the previous chapter. The same rationale applies here, too.

Objective 14.1. Use partial fraction expansion to express the ratio of polynomials as the sum of partial fractions.

The table of transform pairs on page 13.17 is there to avoid the complex integration of Eq. 13.1.2. But the function $F(s)$ must be in one of the forms in the table. If $F(s)$ is a ratio of polynomials in s, then partial fraction expansion can do this. Thus here is a detailed look at partial fraction expansion.

Suppose $F(s)$ is expressed as the ratio of polynomials,

$$F(s) = \frac{a_n s^n + a_{n-1} s^{n-1} + \cdots + a_1 s + a_0}{b_m s^m + b_{m-1} s^{m-1} + \cdots + b_1 s + b_0} = \frac{N(s)}{D(s)} \tag{14.1.1}$$

Then n is the order of the numerator $N(s)$ and m is the order of the denominator $D(s)$. There are two cases: $n \geq m$ and $n < m$. If $n \geq m$ then $F(s)$ is termed an improper fraction. If $n < m$ then the fraction is called proper. If $F(s)$ is improper then it can always be separated into the sum of polynomials in s plus a proper fraction. For example, if

14.1

$$F(s) = \frac{s^3 + 4s^2 + 6s + 6}{s^2 + 3s + 2}$$

Then dividing the denominator into the numerator gives

$$F(s) = s + 1 + \frac{s + 4}{s^2 + 3s + 2}$$

This puts F(s) in the form of a polynomial in positive powers of s plus a proper fraction. We will confine the following discussion to proper fractions.

There are many methods for expanding a proper fraction into partial fractions. Three of these are presented in the following. Which is best? That depends on the situation.

Equating Coefficients

Suppose the proper fraction F(s) is given by

$$F(s) = \frac{N(s)}{(s-s_1)(s-s_2)\cdots(s-s_m)} = \frac{A_1}{s-s_1} + \frac{A_2}{s-s_2} + \cdots + \frac{A_m}{s-s_m} \qquad (14.1.2)$$

The constants A_i are known as residues. Thus A_i is the residue of the pole at $s = s_i$. In most cases the simplest method to evaluate the residues is to equate coefficients. This is the method presented in the following problem.

Problem 14.1.1. Find v(t) if V(s) is given by

$$V(s) = \frac{5}{s\left(s + \frac{1}{2}\right)}, \quad \sigma > 0$$

Solution: There is no transform pair in the table on page 13.16 that matches this function. Therefore partial fraction expansion is necessary. Separate terms to get

14.2

$$V(s) = \frac{5}{s\left(s + \frac{1}{2}\right)} = \frac{A}{s} + \frac{B}{s + \frac{1}{2}}$$

Now combine the two terms on the right side of this equation by finding a common denominator.

$$\frac{A}{s} + \frac{B}{s + \frac{1}{2}} = \frac{A\left(s + \frac{1}{2}\right) + Bs}{s\left(s + \frac{1}{2}\right)}$$

Equate like coefficients in the numerator to solve for the constants A and B. That is, since $A(s + \frac{1}{2}) + Bs = 5$, set

$As + Bs = 0$

and $A/2 = 5$

or $A = 10, B = -10$.

Therefore $V(s)$ is given by

$$V(s) = \frac{10}{s} - \frac{10}{s + 1/2}, \quad \sigma > 0$$

Use the Table of Laplace Transforms on page 13.17 to get

$$v(t) = 10\left(1 - e^{-t/2}\right)u(t)$$

Problem 14.1.2. Find $v(t)$ if $V(s)$ is given by

$$V(s) = \frac{s+1}{s^2 + 4s + 4}, \quad \sigma < -2$$

Solution: The denominator has repeated roots at $s = -2$. Therefore the expansion is given by

$$\frac{s+1}{(s+2)^2} = \frac{A}{s+2} + \frac{B}{(s+2)^2}$$

14.3

Combining terms on the right gives

$$\frac{A}{s+2} + \frac{B}{(s+2)^2} = \frac{A(s+2)+B}{(s+2)^2}$$

Now equate like coefficients in the numerator.

$s + 1 = A(s + 2) + B$

$As = s$ or $A = 1$

$1 = 2A + B$ or $B = -1$

Therefore $V(s)$ is given by

$$V(s) = \frac{1}{s+2} - \frac{1}{(s+2)^2}, \quad \sigma < -2$$

Now use the table to obtain

$$v(t) = \left(-e^{-2t} + te^{-2t}\right)u(-t)$$

Heaviside's Expansion Theorem

If there are a large number of roots, the above method of equating coefficients leads to a large number of simultaneous equations to solve. In this case a simpler procedure, due to Oliver Heaviside (1850-1925), is as follows:

Refer to Eq. 14.1.2. To evaluate A_1 multiply both sides by $(s - s_1)$.

$$\frac{(s - s_1)N(s)}{(s - s_1)(s - s_2)\cdots(s - s_m)} = A_1 + \frac{(s - s_1)A_2}{s - s_2} + \cdots + \frac{(s - s_1)A_m}{s - s_m}$$

This equation is true for every value of s. If $s = s_1$ then every term on the right side is zero except for A_1. The left side is not zero because the $(s - s_1)$ terms cancel. Therefore A_1 is given by

$$A_1 = (s - s_1) \frac{N(s)}{D(s)} \bigg|_{s = s_1}$$

or, in general

$$A_i = (s - s_i) \frac{N(s)}{D(s)} \bigg|_{s = s_i} \qquad (14.1.4)$$

Problem 14.1.3. Find $f(t)$ if $F(s)$ is given by

$$F(s) = \frac{s^4 + 3s^3 + 4s^2 + 2s + 3}{(s+1)(s+2)(s+3)(s+4)(s+5)}$$

Solution: Write

$$\frac{s^4 + 3s^3 + 4s^2 + 2s + 3}{(s+1)(s+2)(s+3)(s+4)(s+5)} = \frac{A_1}{s+1} + \frac{A_2}{s+2} + \frac{A_3}{s+3} + \frac{A_4}{s+4} + \frac{A_5}{s+5}$$

Use Eq. 14.1.4 to obtain

$$A_1 = \frac{s^4 + 3s^3 + 4s^2 + 2s + 3}{(s+2)(s+3)(s+4)(s+5)} \bigg|_{s=-1} = \frac{1}{8}$$

$$A_2 = \frac{s^4 + 3s^3 + 4s^2 + 2s + 3}{(s+1)(s+3)(s+4)(s+5)} \bigg|_{s=-2} = -\frac{7}{6}$$

Similarly,

$$A_3 = \frac{33}{4}, \quad A_4 = -\frac{41}{2}, \quad A_5 = \frac{343}{24}$$

Therefore $F(s)$ is given by

$$F(s) = \frac{1/8}{s+1} - \frac{7/6}{s+2} + \frac{33/4}{s+3} - \frac{41/2}{s+4} + \frac{343/24}{s+5}, \quad -4 < \sigma < -3$$

The table of transforms gives

14.5

$$f(t) = \left[-\frac{1}{8}e^{-t} + \frac{7}{6}e^{-2t} - \frac{33}{4}e^{-3t}\right]u(-t) + \left[-\frac{41}{2}e^{-4t} + \frac{343}{24}e^{-5t}\right]u(t)$$

Note: An attempt to use the method of equating coefficients in this problem should convince you that Heaviside's expansion theorem is worthwhile.

Problem 14.1.4. Find $f(t)$ if $F(s)$ is given by

$$F(s) = \frac{s+2}{s^2 + 2s + 2} = \frac{s+2}{(s+1+j)(s+1-j)}, \quad \sigma > -1$$

Solution: Expand $F(s)$ in partial fractions to obtain

$$\frac{s+2}{(s+1+j)(s+1-j)} = \frac{A}{s+1+j} + \frac{B}{s+1-j}$$

Using Heaviside's method, multiply each term by $s + 1 + j$ to evaluate the residue A.

$$\frac{s+2}{s+1-j} = A + \frac{(s+1+j)B}{s+1-j}$$

or

$$A = \frac{s+2}{s+1-j}\bigg|_{s=-1-j} = 0.5 + j0.5$$

The same procedure will find B, but B must be the conjugate of A, so B is given by

$$B = 0.5 - j0.5$$

Therefore

$$F(s) = \frac{0.5 + j0.5}{s+1+j} + \frac{0.5 - j0.5}{s+1-j}, \quad \sigma > -1$$

Use pair 7 from the table to obtain

14.6

$$f(t) = \left[(0.5 + j0.5)e^{-(1+j)t} + (0.5 - j0.5)e^{-(1-j)t}\right]u(t)$$

Use Euler's formula to put this in a better form.

$$f(t) = e^{-t}\left[\cos t + \sin t\right]u(t)$$

Heaviside's method applies to repeated roots, but this application is not obvious. Consider

$$F(s) = \frac{s}{(s+1)^2} = \frac{A}{(s+1)^2} + \frac{B}{s+1}$$

To find A, multiply both sides by $(s + 1)^2$ to get

$$s = A + (s+1)B \tag{14.1.5}$$

Set $s = -1$ to obtain $A = -1$. But what about B?

The basic idea in algebra is to do the same thing to both sides of an equation. So let us differentiate both sides of Eq. 14.1.5 with respect to s. This gives $B = 1$.

Differentiation is the key. For more than two repeated roots the procedure gets a bit more complicated, so here is an explanation of the procedure: Let

$$\frac{N(s)}{D(s)} = \frac{N(s)}{(s-s_1)^r} = \frac{A_1}{s-s_1} + \cdots + \frac{A_n}{(s-s_1)^n} + \cdots + \frac{A_r}{(s-s_1)^r}$$

Multiply both sides by $(s - s_1)^r$.

$$N(s) = (s-s_1)^{r-1}A_1 + \cdots + (s-s_1)^{r-n}A_n + \cdots + A_r$$

To evaluate A_n differentiate $(r - n)$ times to obtain

$$\frac{d^{r-n}}{ds^{r-n}}N(s) = (r-n)!A_n + (s-s_1)terms$$

14.7

Now set $s = s_1$ to eliminate terms containing $(s - s_1)$. Divide by $(r - n)!$ to obtain the general form given by

$$A_n = \frac{1}{(r-n)!} \frac{d^{r-n}}{ds^{r-n}} N(s)\bigg|_{s = s_1} \quad (14.1.6)$$

In this derivation $D(s)$ consisted entirely of $(s - s_1)^r$. Usually there will be other terms in $D(s)$, so Eq. 14.1.6 must be changed to the form

$$A_n = \frac{1}{(r-n)!} \frac{d^{r-n}}{ds^{r-n}} \left[\frac{N(s)}{D(s)}(s-s_1)^r\right]\bigg|_{s = s_1} \quad (14.1.7)$$

Problem 14.1.4. Use Heaviside's method to find $f(t)$ corresponding to $F(s)$ given by

$$F(s) = \frac{4s^3 + 13s^2 + 14s + 6}{(s+1)^3(s+2)}, \quad \sigma > -1$$

Solution: Expand $F(s)$ in partial fractions.

$$\frac{4s^3 + 13s^2 + 14s + 6}{(s+1)^3(s+2)} =$$

$$\frac{A_1}{(s+1)^3} + \frac{A_2}{(s+1)^2} + \frac{A_3}{s+1} + \frac{A_4}{s+2} \quad (14.1.8)$$

To solve for A_1 multiply both sides by $(s + 1)^3$ to obtain

$$\frac{4s^3 + 13s^2 + 14s + 6}{s+2} =$$

$$A_1 + (s+1)A_2 + (s+1)^2 A_3 + \frac{(s+1)^3}{s+2}A_4 \quad (14.1.9)$$

Set $s = -1$ to evaluate A_1.

$$A_1 = \left. \frac{4s^3 + 13s^2 + 14s + 6}{s+2} \right|_{s=-1} = 1$$

To solve for A_2 differentiate each term in Eq. 14.1.9 with respect to s to obtain

$$\frac{(s+2)(12s^2 + 26s + 14) - (4s^3 + 13s^2 + 14s + 6)}{(s+2)^2} =$$

$$A_2 + 2(s+1)A_3 + \frac{3(s+2)(s+1)^2 - (s+1)^3}{(s+2)^2}A_4$$

(14.1.10)

With $s = -1$ all terms except A_2 are eliminated on the right side. The left side gives $A_2 = -1$.

Take one more derivative to evaluate A_3. We'll leave out the details, but the correct answer is $A_3 = 2$.

Now to evaluate A_4 return to Eq. 14.1.8, multiply both sides by $(s + 2)$ and set $s = -2$. This gives $A_4 = 2$.

Having evaluated all the coefficients we can write

$$F(s) = \frac{1}{(s+1)^3} - \frac{1}{(s+1)^2} + \frac{2}{s+1} + \frac{2}{s+2}, \quad \sigma > -1$$

Use the tables to obtain

$$f(t) = \left[\frac{t^2}{2}e^{-t} - te^{-t} + 2e^{-t} + 2e^{-2t} \right]u(t)$$

Note: This is a complicated problem for any method. Our first method would require the solution to four simultaneous equations, while Heaviside's method required several derivatives. Our third method, presented below, is usually easier to use on complicated problems.

Substitution of Variables

Assume that $F(s)$ has an rth order pole at $s = s_1$. Thus $F(s)$ can be written as

$$F(s) = \frac{N(s)}{D(s)} = \frac{N(s)}{(s-s_1)^r D_1(s)} \qquad (14.1.11)$$

where $D_1(s)$ is the polynomial remaining after $(s - s_1)^r$ has been factored from $D(s)$.

Now multiply both sides of Eq. 14.1.11 by $(s - s_1)^r$ to obtain

$$F(s)(s-s_1)^r = \frac{N(s)}{D_1(s)} \qquad (14.1.12)$$

Make the substitution $p = (s - s_1)$ in Eq. 14.1.12 to obtain

$$F(s)p^r = \frac{N(p+s_1)}{D_1(p+s_1)}$$

It turns out that if, on the right side of this equation, the denominator is divided into the numerator, the result can be used to solve for the r residues of $(s - s_1)^r$. Instead of presenting the procedure formally we will illustrate it with an example.

Refer to Problem 14.1.4. First multiply both sides by $(s + 1)^3$ to obtain

$$F(s)(s+1)^3 = \frac{4s^3 + 13s^2 + 14s + 6}{s+2}$$

Make the substitution $p = s + 1$, i.e., set $s = p - 1$ to obtain

$$F(s)p^3 = \frac{4p^3 + p^2 + 1}{p+1}$$

Now perform long division with both polynomials in ascending powers of p.

14.10

$$\frac{1 - p + 2p^2}{1+p \overline{\smash{\big)}\ 1 + 0 + p^2 + 4p^3}}$$

$$\underline{\quad 1 + p \quad}$$
$$-p + p^2$$
$$\underline{\quad -p - p^2 \quad}$$
$$2p^2 + 4p^3$$
$$\underline{\quad 2p^2 + 2p^3 \quad}$$
$$2p^3$$

Thus
$$F(s)p^3 = 1 - p + 2p^2 + \frac{2p^3}{p+1}$$

Dividing both sides by p^3 and substituting $p = s + 1$ gives

$$F(s) = \frac{1}{(s+1)^3} - \frac{1}{(s+1)^2} + \frac{2}{s+1} + \frac{2}{s+2}, \quad \sigma > -1$$

This completes the partial fraction expansion.

Note: If the last term contained more than the single pole at $s = -2$ this last term would undergo further expansion to complete the solution.

Self Test, Objective 14.1.

Find the inverse transform of $F(s)$ given by

$$F(s) = \frac{-s^2 - 5s - 7}{(s+2)^2(s+1)}, \quad -2 < \sigma < -1$$

Use the method of equating coefficients.

Use Heaviside's method

Use the substitution of variables method.

Objective 14.2. Find the response of an LTI system to an input signal that is expressed by its Laplace transform.

As before with the Fourier series and transform, this objective is the culmination of our previous work on the Laplace transform. There is nothing new here. The procedure is as follows:

1. Find the LTI system transfer function

2. Find the Laplace transform of the input signal.

3. Multiply the input by the transfer function.

4. Find the inverse Laplace transform of the product. This time function is the desired response.

Problem 14.2.1. Use Laplace transforms to find the output voltage in Fig. 14.2.1 if the input is the exponential signal $v_1(t) = e^{-t}u(t)$.

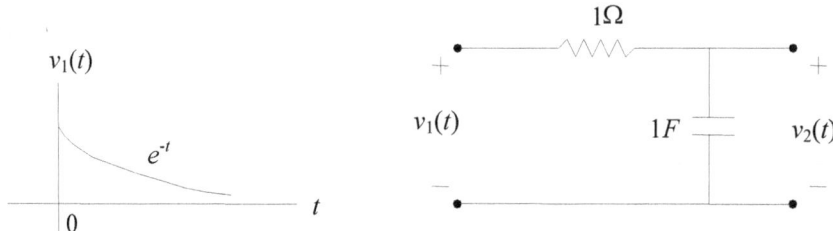

Fig. 14.2.1

Solution: The transfer function is given by

$$H(s) = \frac{1}{s+1}, \quad \sigma > -1$$

14.12

Also
$$V_1(s) = \frac{1}{s+1}, \quad \sigma > -1$$

The product gives $V_2(s)$:

$$V_2(s) = V_1(s)H(s) = \frac{1}{(s+1)^2}, \quad \sigma > -1$$

Therefore

$$v_2(t) = te^{-t}u(t)$$

Problem 14.2.2. Use Laplace transforms to find $i_4(t)$ in Fig. 14.2.2 if $v_1(t)$ is the square pulse signal $v_1(t) = u(t) - u(t-1)$.

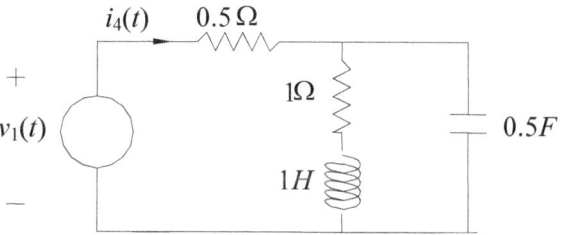

Fig. 14.2.2

Solution: The transfer function is given by

$$H(s) = \frac{2s^2 + 2s + 4}{s^2 + 5s + 6}, \quad \sigma > -2$$

Also

$$V_1(s) = \frac{1}{s} - \frac{1}{s}e^{-s}, \quad -\infty < \sigma < \infty$$

Therefore $I_4(s)$ is given by

$$I_4(s) = \frac{\left(1-e^{-s}\right)\left(2s^2+2s+4\right)}{s(s+3)(s+2)} =$$

$$\frac{2s^2+2s+4}{s(s+3)(s+2)} - \frac{e^{-s}\left(2s^2+2s+4\right)}{s(s+3)(s+2)}, \quad \sigma > -2$$

Since the second term is identical to the first except for the exponential e^{-s}, a bit of thought can save time and effort in finding the inverse transform of this expression. Our strategy is to find the time function corresponding to the first fraction. Then by the time shifting property, the time function for the second term must be the same as the first except for a time shift. Expanding the first term gives

$$\frac{2s^2+2s+4}{s(s+3)(s+2)} = \frac{2/3}{s} + \frac{16/3}{s+3} - \frac{4}{s+4}$$

so that

$$i_4(t) = \left(\tfrac{2}{3} + \tfrac{16}{3}e^{-3t} - 4e^{-2t}\right)u(t)$$
$$- \left(\tfrac{2}{3} + \tfrac{16}{3}e^{-3(t-1)} - 4e^{-2(t-1)}\right)u(t-1)$$

Self Test, Objective 14.2.

Use the Laplace transform to find the response $v_2(t)$ in Fig. 14.2.3 if the input $v_1(t)$ is the square pulse signal $v_1(t) = u(t) - u(t-1)$.

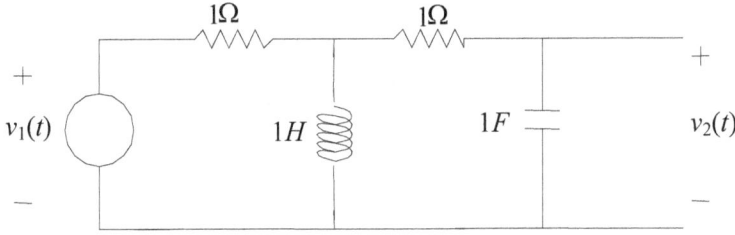

Fig. 14.2.3

Self Test Answers:

Objective 14.1.

$$f(t) = \left[te^{-2t} + 2e^{-2t} \right] u(t) + 3e^{-t} u(-t)$$

Objective 14.2.

$$v_2(t) = \left[e^{-t/2} \sin(t/2) \right] u(t)$$
$$- \left[e^{-(t-1)/2} \sin[(t-1)/2] \right] u(t-1)$$

Chapter 15

Equivalence of System Models

Objectives: After completing this chapter you should be able to do the following:

15.1. Define controllability and observability.
15.2. Determine (select) which systems are controllable and which are observable.
15.3. Derive the equivalent system models (state model, transfer function, and impulse response) for given controllable and observable systems.

Rationale:

This chapter compares the three system models, state model, transfer function, and impulse response. The comparison uses the concepts of controllability and observability.

Controllability implies the ability of the input to influence each state variable. Observability, on the other hand, implies the ability to determine each state variable by observing the output.

R. E. Kalman (1930 -) first introduced these concepts about 1960. Their late arrival is amazing in view of their fundamental nature, for they determine if the transfer function and the impulse response are accurate representations of the system.

Pre -Test, Chapter 15.

Before studying this chapter you should be able to calculate the inverse of the following matrices.

$$A = \begin{bmatrix} 1 & 2 \\ 0 & 1 \end{bmatrix}$$

15.1

$$B = \begin{bmatrix} (s+4) & 2 \\ -1 & (s+1) \end{bmatrix}$$

Objective 15.1. Define controllability and observability.

The concepts of controllability and observability are closely related, so we will discuss both concepts together. First let us review the state variable model for electric networks. Consider the network in Fig. 15.1.1, taken from Chapter 3.

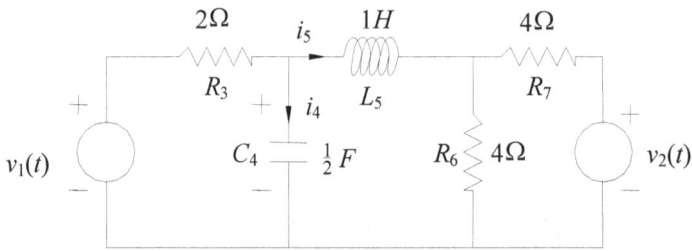

Fig. 15.1.1

From Chapter 3 the state equations are given by

$$\begin{bmatrix} \dot{v}_4 \\ \dot{i}_5 \end{bmatrix} = \begin{bmatrix} -1 & -2 \\ 1 & -2 \end{bmatrix} \begin{bmatrix} v_4 \\ i_5 \end{bmatrix} + \begin{bmatrix} 1 & 0 \\ 0 & -\frac{1}{2} \end{bmatrix} \begin{bmatrix} v_1 \\ v_2 \end{bmatrix} \tag{15.1.1}$$

Recall that one of the distinguishing features of the state variable formulation is that one can find any network variable in terms of the state variables and the forcing function. This means that after solving Eq. 15.1.1 for v_4 and i_5 we could, for example, find the voltage across R_3 and the current through C_4. Let us therefore call $v_3(t)$ and $i_4(t)$ the system output, and denote this output as y.

15.2

$$y = \begin{bmatrix} v_3 \\ i_4 \end{bmatrix}$$

By referring to Fig. 15.1.1 we can use Kirchhoff's laws to express this output vector in terms of the sources v_1, v_2, and the state variables v_4, i_5.

$$v_3 = v_1 - v_4$$

$$i_4 = i_3 - i_5 = \tfrac{v_1}{2} - \tfrac{v_4}{2} - i_5$$

In matrix form these equations become

$$\begin{bmatrix} v_3 \\ i_4 \end{bmatrix} = \begin{bmatrix} -1 & 0 \\ -\frac{1}{2} & -1 \end{bmatrix} \begin{bmatrix} v_4 \\ i_5 \end{bmatrix} + \begin{bmatrix} 1 & 0 \\ \frac{1}{2} & 0 \end{bmatrix} \begin{bmatrix} v_1 \\ v_2 \end{bmatrix} \qquad (15.1.2)$$

Now look at Eqs. 15.1.1 and 15.1.2. They are of the form

$$\dot{x} = Ax + Bq \qquad (15.1.3)$$

$$y = Cx + Dq \qquad (15.1.4)$$

where q is the input, x is the state, and y is the output. Equation 15.1.3 is the state equation and Eq. 15.1.4 is the output equation.

Note: For type B networks the state equation has an additional term involving the derivative of the input. That is, the state equation is in the form

$$\dot{x} = Ax + Bq + E\dot{q} \qquad (15.1.5)$$

This can always be put in the form of Eq. 15.1.3 by the substitution

$$z = x - Eq \qquad (15.1.6)$$

as we did in Chapter 4. Therefore Eq. 15.1.3 is the general form for both type A and type B networks.

With this review you should be ready to tackle the new concepts of controllability and observability.

15.3

Definitions of Controllability and Observability

Uncontrollable is usually bad. If the tie-rods in an automobile come loose, or if the power fails in an airplane, the system becomes uncontrollable. In physical terms, controllability implies that the system is "under control." That is, the inputs to the system are sufficient to change the state from an initial state $x_0 = x(t_0)$ to any specified final state $x_f = x(t_f)$.

Definition 15.1.1. Controllability. A system is said to be controllable if it is possible to find an input $q(t)$ that will transfer the system between two arbitrarily specified finite states x_0 and x_f in finite time $t_f \geq 0$.

Controllability implies the ability of the input to affect each state variable. Observability, on the other hand, implies the ability to determine each state variable by observing the output.

Definition 15.1.2. Observability. A system is said to be observable if every state x_0 can be determined from measurements of the output $y(t)$ over a finite time interval $t_0 \leq t \leq t_f$.

One use, but certainly not the only one, for these definitions is to determine if the transfer function and impulse response adequately represent a system. To see this, consider the partition of a system into four possible subsystems as shown in Fig. 15.1.2.

1. Subsystem S_1 is controllable and observable.
2. Subsystem S_2 is controllable but not observable.
3. Subsystem S_3 is not controllable but is observable.
4. Subsystem S_4 is neither controllable nor observable.

In Fig. 15.1.2 the input q is connected to S_1 and S_2 since they are controllable. The output y is connected to S_1 and S_3 since they are observable.

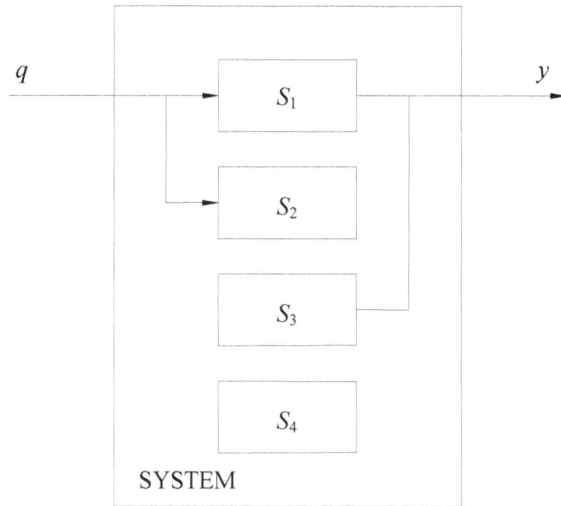

Fig. 15.1.2

It is obvious that the only relationship between input and output is through S_1. Thus the transfer function or impulse response is an accurate representation of only this subsystem. Those portions of the system that are not controllable, or not observable, or both, cannot be represented by the transfer function or impulse response.

Therefore the transfer function is an accurate description of a system if and only if it is both controllable and observable. We can be sure that the transfer function or impulse response representation of a system is equivalent to the state variable model if the system is controllable and observable.

Example 15.1.1. Figure 15.1.3 provides an example of this subsystem concept. The voltage source v_S is the input and y is the output. The state variable formulation is given by

$$\begin{bmatrix} \dot{v}_1 \\ \dot{v}_2 \\ \dot{v}_3 \end{bmatrix} = \begin{bmatrix} -1 & 0 & 0 \\ 0 & -2 & 0 \\ 0 & 0 & -3 \end{bmatrix} \begin{bmatrix} v_1 \\ v_2 \\ v_3 \end{bmatrix} + \begin{bmatrix} 1 \\ 2 \\ 0 \end{bmatrix} v_s \qquad (15.1.7)$$

$$y = \begin{bmatrix} 1 & 0 & 0 \end{bmatrix} \begin{bmatrix} v_1 \\ v_2 \\ v_3 \end{bmatrix} \qquad (15.1.8)$$

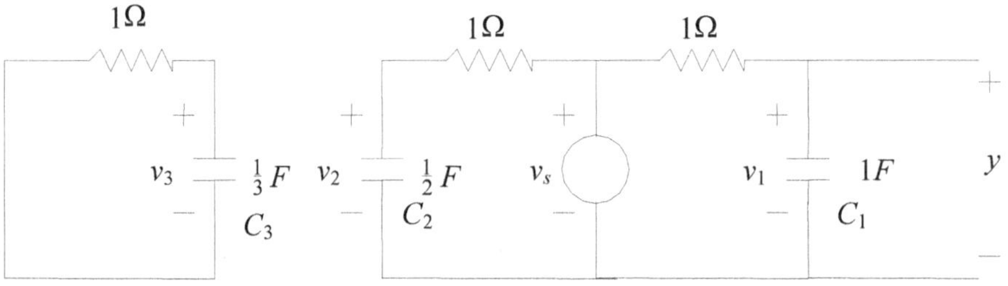

Fig. 15.1.3

It is obvious from the diagram that the input v_s cannot control the value of v_3. Equally obvious is that the output cannot tell us anything about v_2 or v_3. That is, these voltages cannot be determined at some time, say $t = 0$, from a knowledge of $y(t)$ for $0 < t < t_f$.

These same conclusions can be drawn from Eqs. 15.1.7 and 15.1.8. If we write these equations in component form, the third part of Eq. 15.1.7 is given by

$$\dot{v}_3 = -3v_3$$

This equation has no forcing function, i.e., the solution is a function of the initial condition only. Likewise, the output $y(t)$ depends only on $v_1(t)$, so only $v_1(t)$ is observable.

The circuit containing capacitor C_1 is both controllable and observable, and is therefore an example of subsystem S_1. The circuit containing capacitor C_2 is controllable but not observable, an example of S_2. The isolated circuit containing C_3 is neither controllable nor observable, an example of S_4.

The transfer function relating input to output is given by

$$\frac{Y(s)}{V_s(s)} = \frac{1}{s+1}$$

Therefore the transfer function contains less information about the system than does the state variable model.

Example 15.1.2. State variables v_1 and v_2 in Example 15.1.1 are controllable. This implies that we can, by proper selection of the input voltage $v_s(t)$, drive both capacitor voltages from any arbitrary initial values, $v_1(0)$ and $v_2(0)$, to any final values, $v_1(t_f)$ and $v_2(t_f)$, in specified time t_f.

At first it might seem impossible to control two variables with only one input, but this is what controllability implies in this example. In essence, we can drive both capacitor voltages to any value we wish at the end of the time period by proper choice of input voltage.

Suppose we are given the following values:

$v_1(0) = 1$ volt
$v_2(0) = -1$ volt
$v_1(t_f) = -1$ volt
$v_2(t_f) = 1$ volt
$t_f = 1$ second

In other words, drive the voltage across C_1 from +1 volt to –1 volt while driving the voltage across C_2 in the opposite direction, from –1 volt to +1 volt, and all in the same one-second interval.

One of many possible control strategies is shown in Fig. 15.1.4. This particular input voltage is constant at the value E_1 for $0 < t < 0.5$, and is then constant at E_2 for the remainder of the one second time interval. Solving the differential equation with this input and the stated initial conditions does indeed closely approximate the desired final values.

15.7

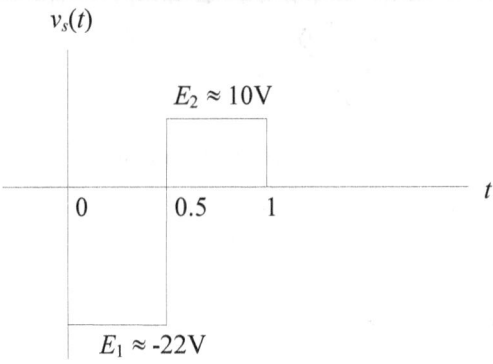

$v_s(t)$

$E_2 \approx 10V$

0   0.5   1 t

$E_1 \approx -22V$

Fig. 15.1.4

Notes: a) An optimum control problem would be to find the function $v_S(t)$ that minimizes the energy dissipated in the resistors while accomplishing the stated goals.

b) In problems of this nature where n state variables must be controlled by one source, one solution is of the form of Fig. 15.1.4. That is, a piecewise constant function that changes value $n - 1$ times in the control interval.

Self Test, Objective 15.1.

1. Write a definition of controllability.

2. Write a definition of observability.

Objective 15.2. Determine (select) which systems are controllable and which are observable.

How can you test the state variable model for controllability and observability? One test for the system described by

15.8

$$\dot{x} = Ax + Bq \qquad\qquad (15.2.1)$$

$$y = Cx + Dq \qquad\qquad (15.2.2)$$

is as follows:

Controllability Test: The system is controllable if and only if the $n{\times}nr$ matrix

$$\left[B \vdots AB \vdots A^2 B \vdots \cdots \vdots A^{n-1} B\right] \qquad\qquad (15.2.3)$$

is of rank n.

Notes: a) The state vector x is an $n{\times}1$ vector, the input q is an $r{\times}1$ vector, the A matrix is $n{\times}n$ and the B matrix is $n{\times}r$.

b) For a single input system ($r = 1$) the system is controllable if and only if the matrix 15.2.3 is nonsingular, i.e., its determinant is nonzero.

c) The rank of a matrix is the number of independent rows or columns in the matrix. (The two are always the same.)

Problem 15.2.1. Test the system in Fig. 15.1.1 for controllability.

Solution: From Eq. 15.1.1 the A and AB matrices are

$$A = \begin{bmatrix} -1 & -2 \\ 1 & -2 \end{bmatrix}$$

$$AB = \begin{bmatrix} -1 & -2 \\ 1 & -2 \end{bmatrix}\begin{bmatrix} 1 & 0 \\ 0 & -\frac{1}{2} \end{bmatrix} = \begin{bmatrix} -1 & 1 \\ 1 & 1 \end{bmatrix}$$

Therefore for our test we have

$$[B \vdots AB] = \begin{bmatrix} 1 & 0 & -1 & 1 \\ 0 & -\frac{1}{2} & 1 & 1 \end{bmatrix}$$

This has rank 2, therefore the system is controllable.

Note: A non-zero matrix has rank n if at least one of its n-square minors is different from zero, while every $(n + 1)$-square minor, if any, is zero. In the above test matrix we can find at least one 2×2 matrix with nonzero determinant.

Problem 15.2.2. Test the system in Fig. 15.1.3 for controllability.

Solution: Refer to Eq. 15.1.7 for the following matrices.

$$B = \begin{bmatrix} 1 \\ 2 \\ 0 \end{bmatrix} \qquad AB = \begin{bmatrix} -1 \\ -4 \\ 0 \end{bmatrix} \qquad A^2B = \begin{bmatrix} 1 \\ 8 \\ 0 \end{bmatrix}$$

The test matrix becomes:

$$\begin{bmatrix} 1 & -1 & 1 \\ 2 & -4 & 8 \\ 0 & 0 & 0 \end{bmatrix}$$

This matrix is singular since it has an all-zero row. Therefore the system is uncontrollable, as we have previously discovered.

Observability Test: A system is observable if and only if the $n \times nm$ composite matrix

$$\left[C^t \vdots A^t C^t \vdots \left(A^t \right)^2 C^t \vdots \cdots \vdots \left(A^t \right)^{n-1} C^t \right] \qquad (15.2.4)$$

is of rank n.

Notes: a) The output vector y is an $m{\times}1$ vector so that C is $m{\times}n$.

b) For a single output system ($m = 1$) the system is observable if and only if the matrix C^t is nonsingular.

Problem 15.2.3. Test the system in Fig. 15.1.1 for observability.

Solution: Form the following matrices from Eqs. 15.1.1 and 2.

$$C' = \begin{bmatrix} -1 & -\frac{1}{2} \\ 0 & 1 \end{bmatrix}$$

$$A'C' = \begin{bmatrix} -1 & 1 \\ -2 & -2 \end{bmatrix}\begin{bmatrix} -1 & -\frac{1}{2} \\ 0 & 1 \end{bmatrix} = \begin{bmatrix} 1 & -\frac{1}{2} \\ 2 & 3 \end{bmatrix}$$

Therefore the test matrix is given by

$$\begin{bmatrix} C' \vdots A'C' \end{bmatrix} = \begin{bmatrix} -1 & -\frac{1}{2} & 1 & -\frac{1}{2} \\ 0 & -1 & 2 & 3 \end{bmatrix}$$

This has rank 2 so the system is observable.

Problem 15.2.4. Test the system in Fig. 15.1.3 for observability.

Solution: Form the test matrix from Eqs. 15.1.7 and 8.

$$C' = \begin{bmatrix} 1 \\ 0 \\ 0 \end{bmatrix} \quad A'C' = \begin{bmatrix} -1 \\ 0 \\ 0 \end{bmatrix} \quad (A')^2 C' = \begin{bmatrix} 1 \\ 0 \\ 0 \end{bmatrix}$$

Combine these three matrices to form the test matrix, given by

$$\left[C' \vdots A'C' \vdots (A')^2 C'\right] = \begin{bmatrix} 1 & -1 & 1 \\ 0 & 0 & 0 \\ 0 & 0 & 0 \end{bmatrix}$$

This matrix is singular so the system is unobservable.

Self Test, Objective 15.2. Test the following system for a) controllability and b) observability.

$$\begin{bmatrix} \dot{x}_1 \\ \dot{x}_2 \end{bmatrix} = \begin{bmatrix} 1 & 2 \\ 0 & 2 \end{bmatrix}\begin{bmatrix} x_1 \\ x_2 \end{bmatrix} + \begin{bmatrix} 1 & 0 \\ \frac{1}{2} & 0 \end{bmatrix}\begin{bmatrix} q_1 \\ q_2 \end{bmatrix}$$

$$\begin{bmatrix} y_1 \\ y_2 \\ y_3 \end{bmatrix} = \begin{bmatrix} 1 & 0 \\ 0 & 2 \\ 0 & 1 \end{bmatrix}\begin{bmatrix} x_1 \\ x_2 \end{bmatrix}$$

Objective 15.3. Derive the equivalent system models (state model, transfer function, and impulse response) for given controllable and observable systems.

If an LTI system is both controllable and observable then the transfer function and impulse response are equivalent to the state variable model. This equivalence means that you can derive any one of the three models from any other model. You should be able to do just that after completing this objective.

First of all, we have applied the concept of transfer function to only single input-single output systems. But it also applies to multiple input-multiple output systems. Consider the system in Fig. 15.1.1. There are two inputs, v_1 and v_2, and two outputs, v_3 and i_4. The relationship between each input and each output creates four transfer functions. In order to find any one of these, set all other sources to zero, i.e., short all other voltage sources and open all other current sources. Then solve for the desired transfer function with all initial conditions set to zero.

There are two ways to find these four transfer functions. The obvious way is to work directly with Fig. 15.1.1 as suggested by the above discussion. But that doesn't help us to understand how to derive the transfer function matrix from the state variable model. Therefore we will derive the relationship between the state model and the transfer function matrix directly. (As an example, this transfer function matrix is, for Fig. 15.1.1, given by)

$$H(s) = \begin{bmatrix} H_{13}(s) & H_{14}(s) \\ H_{23}(s) & H_{24}(s) \end{bmatrix}$$

To begin, here are the state models:

$$\dot{x} = Ax + Bq \tag{15.3.1}$$

$$y = Cx + Dq \tag{15.3.2}$$

Denote the transform of the x, q, and y vectors by $X(s)$, $Q(s)$, and $Y(s)$. That is,

$$X(s) = \begin{bmatrix} X_1(s) \\ X_2(s) \\ \vdots \\ X_n(s) \end{bmatrix} \qquad Q(s) = \begin{bmatrix} Q_1(s) \\ Q_2(s) \\ \vdots \\ Q_r(s) \end{bmatrix} \qquad Y(s) = \begin{bmatrix} Y_1(s) \\ Y_2(s) \\ \vdots \\ Y_m(s) \end{bmatrix}$$

Take the transform of the state models in Eqs. 15.3.1and 2.

$$sX(s) - x(0) = AX(s) + BQ(s) \tag{15.3.3}$$

$$Y(s) = CX(s) + DQ(s) \tag{15.3.4}$$

Now an intermediate result must be derived to use in our discussion. Write Eq. 15.3.3 in the form

$$sX(s) - AX(s) = x(0) + BQ(s)$$

or $\qquad (sI - A)X(s) = x(0) + BQ(s)$

Now pre-multiply by $(sI - A)^{-1}$ to solve for $X(s)$.

15.13

$$X(s) = (sI - A)^{-1}x(0) + (sI - A)^{-1}BQ(s) \qquad (15.3.5)$$

Recall that both the impulse response and transfer function are found with initial conditions equal to zero. Set $x(0) = 0$ to get

$$X(s) = (sI - A)^{-1}BQ(s)$$

Substitute this into the output equation, Eq. 15.3.4, to get

$$Y(s) = C(sI - A)^{-1}BQ(s) + DQ(s)$$

Compare this to

$$Y(s) = H(s)Q(s)$$

to arrive at the following expression for $H(s)$.

$$H(s) = C(sI - A)^{-1}B + D \qquad (15.3.7)$$

This leads to $h(t)$ by the inverse transform.

$$h(t) = L^{-1}H(s) = Ce^{At}B + D \qquad (15.3.8)$$

where L^{-1} stands for the inverse Laplace transform and the matrix exponential e^{At} has the transform

$$e^{At} \leftrightarrow (sI - A)^{-1} \qquad (15.3.9)$$

Of fundamental importance, this transform is derived in most system textbooks. Despite its importance, we simply have not needed it before, and this is not the place to derive it. Just add this to the table of transform pairs.

Problem 15.3.1. Find the transfer function and impulse response matrix of Fig. 15.1.1.

Solution:

$$(sI - A) = \begin{bmatrix} s+1 & 2 \\ -1 & s+2 \end{bmatrix}$$

The inverse is given by

$$(sI - A)^{-1} = \frac{1}{s^2 + 3s + 4} \begin{bmatrix} s+2 & -2 \\ 1 & s+1 \end{bmatrix}$$

Therefore the transfer function matrix is given by

$$H(s) = C(sI - A)^{-1}B + D$$

$$= \frac{1}{s^2 + 3s + 4} \begin{bmatrix} (s^2 + 2s + 2) & -1 \\ \frac{1}{2}(s^2 + 2s) & \frac{1}{2}s \end{bmatrix}$$

Take the inverse transform to find the impulse response matrix $h(t)$.

$$h(t) = \begin{bmatrix} h_{13}(t) & h_{14}(t) \\ h_{23}(t) & h_{24}(t) \end{bmatrix}$$

where

$$h_{13}(t) = \delta(t) - e^{-3t}\left[\cos\frac{\sqrt{7}}{2}t + \frac{1}{\sqrt{7}}\sin\frac{\sqrt{7}}{2}t\right]u(t)$$

$$h_{14}(t) = -\left[\frac{2}{\sqrt{7}}e^{-\frac{3}{2}t}\sin\frac{\sqrt{7}}{2}t\right]u(t)$$

$$h_{23}(t) = \frac{1}{2}\delta(t) - e^{-\frac{3}{2}t}\left[\frac{1}{2}\cos\frac{\sqrt{7}}{2}t + \frac{5}{2\sqrt{7}}\sin\frac{\sqrt{7}}{2}t\right]u(t)$$

$$h_{24}(t) = e^{-\frac{3}{2}t}\left[\frac{1}{2}\cos\frac{\sqrt{7}}{2}t - \frac{3}{2\sqrt{7}}\sin\frac{\sqrt{7}}{2}t\right]u(t)$$

15.15

Problem 15.3.2. Consider the circuit shown in Fig. 15.3.1. There is one input $v_1(t)$. Suppose we are interested in finding the two currents $i_2(t)$ and $i_4(t)$. Do the following:

a) Write the system state and output equations and test the system for controllability and observability. Use $v_2(t)$ and $i_3(t)$ as the state variables.

b) If the system is controllable and observable, compute the transfer function matrix and impulse response matrix.

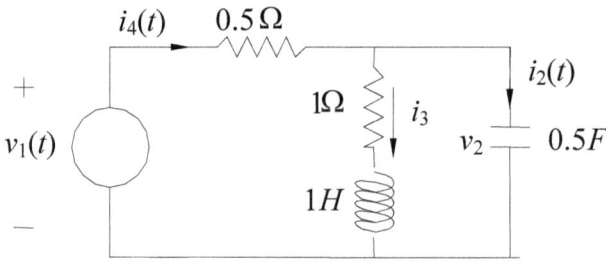

Fig. 15.3.1

Solution: a) The state and output equations are given by

$$\begin{bmatrix} \dot{v}_2 \\ \dot{i}_3 \end{bmatrix} = \begin{bmatrix} -4 & -2 \\ 1 & -1 \end{bmatrix} \begin{bmatrix} v_2 \\ i_3 \end{bmatrix} + \begin{bmatrix} 4 \\ 0 \end{bmatrix} v_1$$

$$\begin{bmatrix} i_2 \\ i_4 \end{bmatrix} = \begin{bmatrix} -2 & -1 \\ -2 & 0 \end{bmatrix} \begin{bmatrix} v_2 \\ i_3 \end{bmatrix} + \begin{bmatrix} 2 \\ 2 \end{bmatrix} v_1$$

Test for controllability:

$$B = \begin{bmatrix} 4 \\ 0 \end{bmatrix}, \quad AB = \begin{bmatrix} -4 & -2 \\ 1 & -1 \end{bmatrix} \begin{bmatrix} 4 \\ 0 \end{bmatrix} = \begin{bmatrix} -16 \\ 4 \end{bmatrix}$$

$$[B \vdots AB] = \begin{bmatrix} 4 & -16 \\ 0 & 4 \end{bmatrix}$$

This has rank 2, so the system is controllable.

Test for observability: $\quad C' = \begin{bmatrix} -2 & -2 \\ -2 & 0 \end{bmatrix}$,

$$A'C' = \begin{bmatrix} -4 & 1 \\ -2 & -1 \end{bmatrix}\begin{bmatrix} -2 & -2 \\ -1 & 0 \end{bmatrix} = \begin{bmatrix} 7 & 8 \\ 5 & 4 \end{bmatrix}$$

$$[C' \vdots A'C'] = \begin{bmatrix} -2 & -2 & 7 & 8 \\ -1 & 0 & 5 & 4 \end{bmatrix}$$

This has rank 2, so the system is observable.

b) To compute the transfer function matrix use Eq. 15.3.7 and proceed as follows:

$$(sI - A) = \begin{bmatrix} s+4 & 2 \\ -1 & s+1 \end{bmatrix}$$

$$(sI - A)^{-1} = \frac{1}{s^2 + 5s + 6}\begin{bmatrix} s+1 & -2 \\ 1 & s+4 \end{bmatrix}$$

$$C(sI - A)^{-1}B + D = \frac{-4}{s^2 + 5s + 6}\begin{bmatrix} 2s+3 \\ 2s+2 \end{bmatrix} + \begin{bmatrix} 2 \\ 2 \end{bmatrix}$$

or

$$H(s) = \frac{2}{s^2 + 5s + 6}\begin{bmatrix} s(s+1) \\ s^2 + s + 2 \end{bmatrix}$$

This is in the form

15.17

$$H(s) = \begin{bmatrix} H_{12}(s) \\ H_{14}(s) \end{bmatrix}$$

Where H_{12} is the relationship between v_1 and i_2, and H_{14} is the relationship between v_1 and i_4. The inverse transform gives

$$h(t) = \begin{bmatrix} 2\delta(t) - 12e^{-3t} + 4e^{-2t} \\ 2\delta(t) - 16e^{-3t} + 8e^{-2t} \end{bmatrix} u(t)$$

Self Test, Objective 15.3.

The circuit in Fig. 15.3.2 has one input $v_1(t)$ and two outputs of interest, $v_2(t)$ and $v_5(t)$.

a) Derive the state model and test for controllability and observability.

b) If possible, find the equivalent transfer function and impulse response matrices.

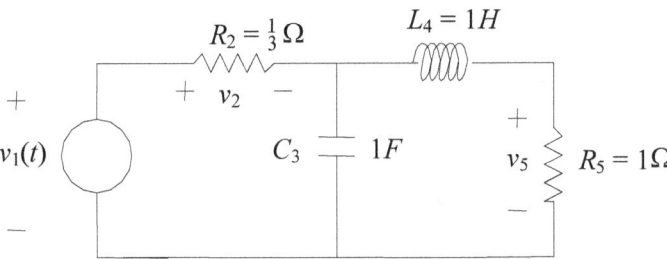

Fig. 15.3.2

Pre Test Answers:

$$A^{-1} = \begin{bmatrix} 1 & -2 \\ 0 & 1 \end{bmatrix}$$

$$B^{-1} = \frac{1}{s^2 + 5s + 6} \begin{bmatrix} (s+1) & -2 \\ 1 & (s+4) \end{bmatrix}$$

Self Test Answers:

Objective 15.1

See definitions 15.1.1 and 15.1.2.

Objective 15.2

    a) No, the system is not controllable.
    b) Yes, the system is observable.

Objective 15.3.

$$\begin{bmatrix} \dot{v}_3 \\ \dot{i}_4 \end{bmatrix} = \begin{bmatrix} -3 & -1 \\ 1 & -1 \end{bmatrix} \begin{bmatrix} v_3 \\ i_4 \end{bmatrix} + \begin{bmatrix} 3 \\ 0 \end{bmatrix} v_1$$

$$\begin{bmatrix} v_2 \\ v_5 \end{bmatrix} = \begin{bmatrix} -1 & 0 \\ 0 & 1 \end{bmatrix} \begin{bmatrix} v_3 \\ i_4 \end{bmatrix} + \begin{bmatrix} 1 \\ 0 \end{bmatrix} v_1$$

The system is both controllable and observable.

$$(sI - A)^{-1} = \frac{1}{s^2 + 4s + 4} \begin{bmatrix} s+1 & -1 \\ 1 & s+3 \end{bmatrix}$$

15.19

$$H(s) = \frac{1}{s^2 + 4s + 4} \begin{bmatrix} s^2 + s + 1 \\ 3 \end{bmatrix}$$

$$h(t) = \begin{bmatrix} \delta(t) - 3e^{-2t} + 3te^{-2t} \\ 3te^{-2t} \end{bmatrix} u(t)$$

INDEX

Technical LAP Series:

Volume 1 in the Technical LAP Series. A description of mathematical concepts in plain language. Vector spaces, symbolic derivative and steepest descent, matrix of transformation, least squares and the pseudo inverse, probability and random variables, eigenvectors, crisp and fuzzy logic, and entropy. These concepts are first defined and then illustrated with engineering applications to add context.

Volume 2 in the Technical LAP Series. What is entropy? What is a fuzzy system? How can one use these in pattern recognition?
What can artificial neural networks do? How can one reduce dimensions with a minimum loss of information, or design templates that take into account the similarities and differences between classes?

CONTINUOUS-TIME
TRANSFORMS FOR ENGINEERS
AND SCIENTISTS

Dwight F Mix

Volume 3 in the Technical LAP Series. Aimed at students, practicing engineers, and scientists, this text presents unique insight into the continuous-time Fourier transform and series. How to plot complex exponential signals. How to use this technique to calculate the transform at a specific frequency by finding the center of mass in the complex plane. Three chapters (LAPS) are devoted to the Laplace transform, one on the forward transform, one on the inverse transform, and one on properties.

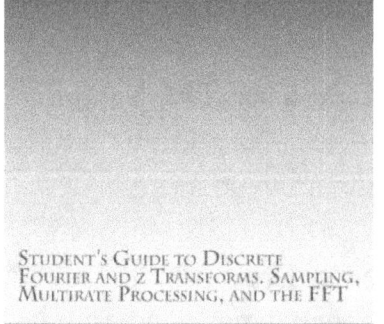

STUDENT'S GUIDE TO DISCRETE
FOURIER AND z TRANSFORMS, SAMPLING,
MULTIRATE PROCESSING, AND THE FFT

Dwight F. Mix

Volume 4 in the Technical LAP Series. For students, practicing engineers, and scientists. Discrete-time Fourier transforms, z transforms, sampling, multi-rate processing, and the fast Fourier transform. An introduction to the pulse sorting transform. This is a unique modification to the Fourier transform for the purpose of sorting pulse signals.

Student's Guide to Fourier, Laplace, and z Transcorms

Technical LAP Series, Vol. 5

Dwight F. Mix

Volume 5 in the Technical LAP Series combines Volumes 3 and 4 into one volume. After showing how to plot complex exponential signals, this text introduces all four forms of the Fourier transform. The forward and inverse Laplace transform for continuous-time signals, and the forward and inverse z transform for discrete-time signals. Insight into the process of finding transforms. Specifically how to estimate the Fourier transform of both continuous-time and discrete-time signals from Argand plots of complex exponential signals.

Student's Guide to Analog Systems.

State Equations, Transforms, Convolution, Controllability and Observability

Dwight F Mix

The three methods for finding the response of an LTI system are differential equations, transform methods, and convolution. Under the conditions of controllability and observability these three methods are equivalent. If the system is not LTI there are no general methods for finding the response. This text introduces the LTI properties plus controllability and observability, and shows their connection to all three methods.

Filter Design Techniques

Analog and Digital Filters, Frequency Selective and Matched
Filters, Adaptive Matched Filter and Template Design

Dwight F. Mix

Filter Design Techniques explains how to design analog, digital, and matched filters. It is intended for practicing engineers and scientists who have a background in Fourier, Laplace, and z transforms.

Part 1 is concerned with analog Butterworth and Chebyshev filter design. Part 2 explains IIR and FIR digital filter design. Part 3 introduces adaptive design of templates for pattern recognition and matched filters for signal detection.

The design technique in Part 3 takes into account both the signal to be detected and the differences and similarities to all the other signals or patterns of interest.